D0581982

ENERGY MANAGERS' *HANDBOOK*

Graham & Trotman

Published in 1985 by

Graham & Trotman Limited
Sterling House
66 Wilton Road
London SW1V 1DE

Head Office:
NIFES House
Sinderland Road
Altrincham
Cheshire WA14 5HQ
Tel. 061 928 5791

ISBN 0 86010 619 5

Printed in Great Britain by
Thetford Press Limited, Thetford, Norfolk.
Bound by Standard Bookbinders, London

CONTENTS

FOREWORD

In recent years the concept of the Energy Manager has been encouraged and developed by the Department of Energy, particularly through its monthly newspaper "Energy Management". There must be few firms or commercial enterprises who have not appointed one of their staff to be such an Energy Manager. Today there is no real shortage of energy supplies, so that conservation is not an immediate necessity in order to help bridge any supply gap. However, there is a finite reserve of fossil fuels, so conservation today can extend the future life of these and give more time for alternatives, such as safe nuclear fusion or solar power, to be developed. More immediately, efficient use of energy can reduce costs and so help industry to save money, be more competitive, or create funds for profit or investment.

Energy Managers, or any other engineer or manager concerned with efficiency, have to cover a very wide range of energy usage and need a source of quick reference to enable them to make reasonable appraisals of problems arising each day. It is hoped that this Handbook will fulfil this need. It comprises a wide range of data, tables, conversions, formulae and notes which should prove helpful. Most of this material can, of course, be found in textbooks, literature issued by manufacturers, papers in technical and professional journals or government bulletins and legislation. However, it is not usually possible to have such a large library at the side of the desk, or to know quickly where to look for the piece of data needed to help solve today's problem. This handbook does not duplicate directly any of this information and indeed most of the diagrams and Tables have been produced within the offices of NIFES, originally for internal use and to assist new junior staff, and also intended for practical use in solving problems rather than to teach theory.

Figures are nearly all presented in SI Units, but where possible without unduly complicating tables, etc., the older "imperial" units are also shown, as many older (and more experienced?) engineers still think more easily about energy problems in the units of their youth. This present version of the Handbook follows on from the success of the earlier Energy Users' Databook, which was published in 1981, but it has been completely rewritten, not just revised or updated.

Improving efficiency is not a once-off exercise on which a few weeks or months are spent and then the exercise can be terminated. Targets should be set for each department or if possible each large energy user, and these should be checked regularly. Energy Managers should not hesitate to ask for advice — there are several Government schemes currently available to help with the cost. Ideas rejected some years ago should be re-examined since energy costs have generally increased at a greater rate than installation costs of new equipment and also new techniques and improved designs may well improve viability.

The energy efficiency campaign never ends. Changes in the work force, changes in materials, new manufacturing techniques and new product ranges all occur and must be compensated for in the energy usage of the plant. Labour must be re-educated and their interest maintained, as otherwise the gratifying results initially obtained will slowly be eroded and costs will start to rise again.

W. SHORT
NIFES

1. UNITS

Metrication, SI Units and Imperial Measure

The conversion period for the bringing of SI Units into general use in Britain should have been completed in 1975. However, it still drags on and many imperial units of measure are still in common use and quoted in reports, catalogues and drawings. Consequently it is still necessary to include pages in tabular form giving the multiplying factors appropriate in converting from Imperial to SI or vice versa, and these follow on the next pages.

These tables should be used as follows:

(a) To convert an Imperial unit to SI, find the appropriate unit in the first (left-hand) column of the Table, read off the multiplying factor in the second column and use this to convert the quantity in Imperial to the required SI equivalent.

(b) To convert an SI unit to Imperial, find the appropriate unit in the third (centre) column of the Table, read off the multiplying factor in the fourth column and use this to convert the SI quantity to Imperial.

The tables include a few units, such as US gallons and barrels, and kilocalories, which are still often met in books and old reports but which are neither Imperial nor SI, and suitable conversion factors are given for these.

Other Derived SI Units

Force	1 Newton (N)	= 1 kgf m/sec^2
Pressure	1 bar	= 10^5 N/m^2
Energy	1 Joule (J)	= 1 Nm
Mass	1 Tonne	= 10^3 kg
Volume	1 Litre (l)	= 10^{-3} m^3
Temperature	Celsius (°C)	= K – 273.15
Thermal Power	1 Watt (W)	= 1 J/sec
Mechanical Power	1 Watt (W)	= 1 Nm/sec
Electrical Power	1 Watt (W)	= 1 VA

Multiples and Sub-multiples of SI Units

Prefix	*Symbol*	*Factor*
Micro	μ	10^{-6}
Milli	m	10^{-3}
Unit	—	1
Kilo	k	10^3
Mega	M	10^6
Giga	G	10^9
Tera	T	10^{12}

Unit Conversion Table

Imperial Unit	× *Factor*	*SI Unit*	× *Factor*	*Imperial Unit*
Mass				
grains (7000 = 1lb)	× 0.06480	g (gramme)	× 15.43	grains
oz (ounce)	× 28.35	g	× 0.03528	oz
lb (pound)	× 453.6	g	× 0.002205	lb
lb	× 0.4536	kg	× 2.205	lb
ton (2240lb)	× 1.016	tonne (1000kg)	× 0.9842	ton

1. UNITS

Imperial Unit	× *Factor*	*SI Unit*	× *Factor*	*Imperial Unit*
Length				
in (inches)	× 25.40	mm	× 0.03937	in
ft (feet)	× 0.3048	m (metres)	× 3.281	ft
miles	× 1.609	km	× 0.6214	miles
Area				
in^2	× 645.2	mm^2	× 0.00155	in^2
in^2	× 0.000645	m^2	× 1550	in^2
ft^2	× 0.0929	m^2	× 10.76	ft^2
yd^2	× 0.8361	m^2	× 1.196	yd^2
acre (43560ft^2)	× 0.4046	Hectare (10,000m^2)	× 2.4716	acre
Volume				
gal (Imperial)	× 4.546	1 (litres)	× 0.220	gal (imp)
gal (Imperial)	× 0.004546	m^3	× 220	gal (imp)
gal (US=0.8327 imp.gal)	× 3.785	1 (litres)	× 0.1832	gal (US)
US barrel	× 159	1 (litres)	× 0.00629	US barrel (35imp.gal)
ft^3	× 28.32	1	× 0.03531	ft^3
ft^3	× 0.02832	m^3	× 35.31	ft^3
yd^3	× 0.7646	m^3	× 1.308	yd^3
Velocity				
ft/sec	× 0.3048	m/sec	× 3.281	ft/sec
mph	× 26.82	m/sec	× 0.0373	mph
mph	× 1.609	km/h	× 0.6214	mph
Flow Rate				
cfm (ft^3/min)	× 0.004719	m^3/sec	× 2119	cfm
cfm	× 0.4719	l/sec	× 2.119	cfm
gal/min (imp)	× 0.2728	m^3/h	× 3.666	gal/min
gal/min (imp)	× 0.07577	l/sec	× 13.20	gal/min
Energy				
Btu	× 1.055	kJ	× 0.9478	Btu
therms (100,000 Btu)	× 105.5	MJ	× 0.009478	therms
Btu	× 0.0002931	kWh	× 3412	Btu
therms	× 29.31	kWh	× 0.03412	therms
kWh	× 3.60	MJ	× 0.2778	kWh
hph (horse power/hr)	× 2.684	MJ	× 0.3725	hph
(also kcal	× 4.187	kJ	× 0.2398	kcal
Btu	× 0.252	k cal	× 3.968	Btu)
Power, heat or energy flow rate				
Btu/h	× 0.0002931	kW	× 3412	Btu/h
therms/h	× 29.31	kW	× 0.03412	therms/h
hp	× 0.7457	kW	× 1.341	hp
hp	× 748	J/sec (or W)	× 0.00134	hp
tons refrigeration	× 3.517	kW	× 0.2843	tons ref. (=12000 Btu/h)
steam lb/h	× 0.2844	kW	× 3.517	steam lb/h
steam kg/h	× 0.6269	kW	× 1.595	steam kg/h
(Note: Steam conditions from and at 100 °C (212 °F), i.e. 970.3 Btu/lb or 2257 kJ/kg)				
Specific heat				
Btu/lb °F	× 4.187	kJ/kg °C	× 0.2388	Btu/lb °F
Btu/ft^3 °F	× 67.07	kJ/m^3 °C	× 0.01491	Btu/ft^3 °F
Pressure				
psi (lb f/in^2)	× 0.06895	bar (100kPa)	× 14.50	psi
psi	× 6.895	kPa	× 0.1450	psi
psi	× 68.95	mbar	× 0.01450	psi
psi	× 703.1	mm H_20	× 0.001422	psi
in Hg (0.4912 psi)	× 33.86	mbar	× 0.02953	in Hg
in H_20 (0.0361 psi)	× 2.491	mbar	× 0.4015	in H_20
standard atmosphere (29.91 in Hg or 14.70psi)	× 1.013	bar	× 0.9872	Standard atmosphere (760mm 1.013 bar)

1. UNITS

Imperial Unit	× *Factor*	*SI Unit*	× *Factor*	*Imperial Unit*
Note: 1mm Hg = 1.333 mbar; 1mm H_20 = 0.09807 mbar. 1kg/cm^2 = 14.22 psi or 0.98 bar.				
Calorific value, heat content				
Btu/lb	× 2.326	kJ/kg (MJ/Tonne)	× 0.4299	Btu/lb
Btu/lb	× 0.002326	MJ/kg (GJ/Tonne)	× 429.9	Btu/lb
Btu/ft^3	× 37.26	kJ/m^3	× 0.02684	Btu/ft^3
Btu/ft^3	× 0.03726	kJ/l (MJ/m^3)	× 26.84	Btu/ft^3
Btu/gallon	× 0.2321	kJ/l (MJ/m^3)	× 4.309	Btu/gal
therms/gal	× 23.21	MJ/l (GJ/m^3)	× 0.04309	therms/gal
(also kcal/kg	× 4.187	kJ/kg	× 0.2388	kcal/kg
Btu/lb	× 0.5556	kcal/kg	× 1.80	Btu/lb
Btu/lb	× 0.0224	therms/ton	× 44.64	Btu/lb
Btu/lb	× 0.0220	therms/tonne	× 45.45	Btu/lb)
Thermal conductivity				
Btu in/ft^{2}h °F	× 0.1442	W/m °C	× 6.933	Btu in/ft^{2}h °F
Heat transfer coefficient				
Btu/ft^{2}h °F	× 5.678	W/m^2 °C	× 0.1761	Btu/ft^{2}h °F
Heat transfer				
Btu/ft^{2}h	× 0.003155	kW/m^2	× 317.0	Btu/ft^{2}h
Btu/ft^{2}h	× 3.155	W/m^2	× 0.317	Btu/ft^{2}h
Density, concentration, humidity				
lb/ft^3	× 0.01602	kg/l	× 62.43	lb/ft^3
lb/ft^3	× 16.02	kg/m^3	× 0.06243	lb/ft^3
lb/gal	× 0.09978	kg/l	× 10.2	lb/gal
grains/ft^3 (7000gr = 1lb)	× 0.002288	kg/m^3	× 437.0	grains/ft^3
grains/lb	× 0.000143	kg/kg	× 7000	grains/lb
Costs				
£/therm	× 9.478	£/GJ	× 0.1055	£/therm
£/lb	× 2.205	£/kg	× 0.4536	£/lb
£/ton	× 0.9842	£/tonne	× 1.016	£/ton
£/foot	× 3.281	£/m	× 0.3048	£/foot
£/ft^2	× 10.76	£/m^2	× 0.0929	£/ft^2
£/ft^3	× 0.03531	£/l	× 28.32	£/ft^3
£/ft^3	× 35.31	£/m^3	× 0.02832	£/ft^3
£/gal (imp)	× 0.2200	£/l	× 4.546	£/gal (imp)

Diagram to Obtain Powers or Roots of Numbers

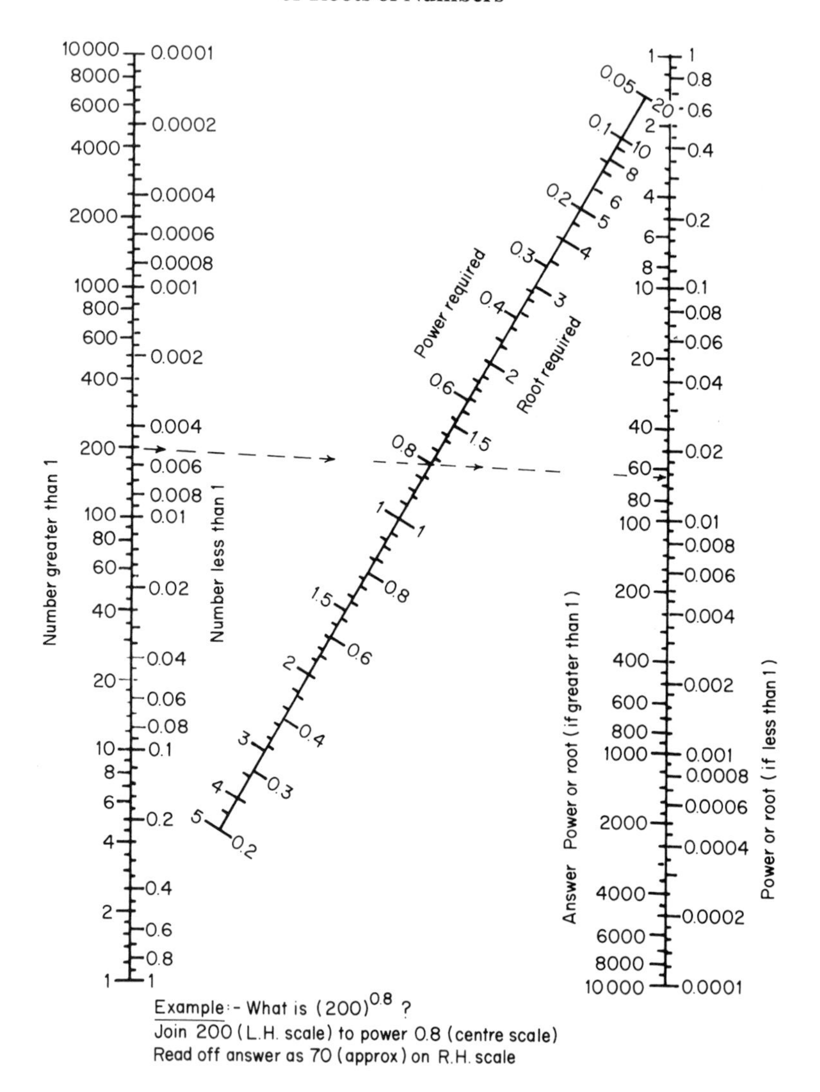

Example:- What is $(200)^{0.8}$?
Join 200 (L.H. scale) to power 0.8 (centre scale)
Read off answer as 70 (approx) on R.H. scale

Temperature Conversion Table

°F		°C	°F		°C	°F		°C
−58	←−50→	−45.6	64.4	←18→	−7.8	140.0	←60→	15.6
−49	−45	−42.8	68.0	20	−6.7	143.6	62	16.7
−40	−40	−40.0	71.6	22	−5.6	147.2	64	17.8
−31	−35	−37.2	75.2	24	−4.4	150.8	66	18.9
−22	−30	−34.4	78.8	26	−3.3	154.4	68	20.0
−13	−25	−31.6	82.4	28	−2.2	158.0	70	21.1
−4	−20	−28.9	86.0	30	−1.1	161.6	72	22.2
5	−15	−26.1						
14	−10	−23.3	89.6	32	0	165.2	74	23.3
17.6	−8	−22.2	93.2	34	1.1	168.8	76	24.4
21.2	−6	−31.1	96.8	36	2.2	172.4	78	25.6
24.8	−4	−20.0	100.4	38	3.3	176.0	80	26.7
28.4	−2	−18.9	104.0	40	4.4	179.6	82	27.8
32.0	0	−17.8	107.6	42	5.6	183.2	84	28.9
35.6	2	−16.7	111.2	44	6.7	186.8	86	30.0
39.2	4	−15.6	114.8	46	7.8	190.4	88	31.1
42.8	6	−14.4	118.4	48	8.9	194.0	90	32.2
46.4	8	−13.3	122.0	50	10.0	197.6	92	33.3
50.0	10	−12.2	125.6	52	11.1	201.1	94	34.4
53.6	12	−11.1	129.2	54	12.2	204.8	96	35.6
57.2	14	−10.0	132.8	56	13.3	208.4	98	36.7
60.8	16	−8.9	136.4	58	14.4	212.0	100	37.8

If temperature is above 100 units (°F or °C), then look up the conversion figure for the last two figures from the above Table for up to 100 units, and add to this the following values for the hundreds.

Add to F		*Add to C*	*Add to F*		*Add to C*
180	←100→	55.5	1980	←1100→	611.1
360	200	111.5	2160	1200	666.7
540	300	166.7	2340	1300	722.2
720	400	222.2	2520	1400	777.7
900	500	277.8	2700	1500	833.3
1080	600	333.3	2880	1600	888.9
1260	700	388.9	3060	1700	944.4
1440	800	444.4	3240	1800	1000
1620	900	500	3420	1900	1055.5
1800	1000	555.5			

This second table is not a straight conversion chart to °F or °C

Examples 1. What is 188°C converted to °F?

Look up 88 in centre column of first table	=	190.4°F
Look up 100 in second table	=	180
Answer 188°C		370.4°F

2. What is 1260°F converted to °C?

Look up 60 in centre column of first table	=	140.0°C
Look up 1200 in second table	=	666.7
Answer 1260 °F		806.7°C

1. UNITS

Formulae Relating to Area and Volume

Surface Areas

Circle	(3 1/7 or 3.1416 × radius × radius i.e. πr^2.)
Triangle	½ × base × perpendicular height
Rectangle	length × breadth
Parallelogram	base × perpendicular height
Cone	½ slant height × perimeter of base + area of base
Sphere	$\pi \times (\text{diameter})^2$
Ellipse	product of axes × 0.7854
Cylinder	2π × radius × height (curved surface) + 2π $(\text{radius})^2$ (two ends)
Pyramid	½ × slant height × total length of all sides at base
Regular Polygon	Number of sides × length of a side × (½ radius of a circle drawn inside polygon that touches centre of sides)
Parabolic Segment	⅔ (diameter of flat end × vertical height)
Annular Ring	¼π (external diameter^2 – internal diameter^2)
Sector of Circle	½ (Radius × length of curved portion) *or* π × Radius^2 × sector Angle/360. (If curved perimeter and radius are known, sector angle = (57.296 × curved perimeter)/radius)
Segment of Circle	½ (radius × length of curved perimeter) – length of arc × (radius – vertical height of segment)
Segment of Sphere	2π (radius of sphere × vertical height of segment) gives curved surface area.
Sector of Sphere	π(radius of sphere) × (2 × vertical height of curved part + diameter of sector/3)

Volumes

Rectangle	length × breadth × height
Cylinder	area of base × height
Prism	area of base × height
Cone	area of base × ⅓ × perpendicular height
Pyramid	area of base × ⅓ × perpendicular height
Sphere	π × ⁴⁄₃ $(\text{radius})^3$
Segment of Sphere	π/24 × vertical height of segment (3 × (diameter of base of $\text{segment})^3$ + 4 × (vertical height of $\text{segment})^2$
Sector of Sphere	$2\pi(\text{radius of sphere})^2$ (vertical height of sector/3)

Properties of Steam

The following tables show values of sensible, latent and total heats of steam for various pressures. These have been rounded off to the nearest unit and will be accurate enough for calculations of boiler efficiencies, flash steam losses and steam needed to affect heat losses, particularly in view of the other likely errors in measurements. While the intervals between pressures may seem rather large, linear interpolation will give quite accurate values for intermediate pressures. If greater accuracy is required, more detailed steam tables in text books, or published as separate booklets, should be studied.

The terms latent heat, sensible heat and total heat have been used as these are probably more familiar to engineers. However, newer titles (for exactly the same figures) are specific enthalpy of evaporation, specific enthalpy of water and specific enthalpy of steam.

For pressures below the standard atmosphere condition of 1.013 bar (14.69 lb/m^2) absolute pressures are used, but at or above atmospheric pressure, gauge pressures are used as being the more likely to be useful for everyday problems.

Absolute Pressure	*Saturation Temp. °C*	*Steam Volume*	*Sensible Heat*	*Latent Heat*	*Total Heat*	*Total Heat, if Superheated to a Final Temperature In °C as below (kJ/kg)*					
Bar		*m^3/kg*	*kJ/kg*	*kJ/kg*	*kJ/kg*	*100*	*150*	*200*	*250*	*300*	*350*
0.05	32.9	28.2	138	2424	2562	2688	2784	2880	2978	3077	3178
0.10	45.8	14.7	192	2393	2585	2688	2783	2880	2977	3077	3178
0.20	60.0	7.65	251	2358	2610	2687	2783	2879	2976	3076	3178
0.40	75.9	3.99	318	2319	2637	2685	2782	2878	2976	3076	3178
0.60	85.9	2.73	360	2294	2654	2681	2781	2877	2975	3076	3177
0.80	93.5	2.09	392	2274	2666	2678	2778	2876	2975	3075	3177
1.00	99.6	1.69	417	2258	2675	2676	2777	2876	2975	3075	3177

Properties of Steam

Gauge Pressure	*Saturation Temp. °C*	*Steam Volume*	*Sensible Heat*	*Latent Heat*	*Total Heat*	*Total Heat, if Superheated to a Final Temperature In °C as below (kJ/kg)*					
Bar		*m^3/kg*	*kJ/kg*	*kJ/kg*	*kJ/kg*	*150*	*200*	*250*	*300*	*350*	*400*
0	100.0	1.67	419	2257	2676	2777	2876	2975	3075	3177	3278
0.2	105.1	1.41	441	2243	2684	2775	2875	2974	3075	3176	3278
0.4	109.6	1.23	460	2231	2691	2774	2874	2974	3074	3176	3278
0.6	113.6	1.08	476	2220	2697	2772	2873	2973	3073	3175	3277
0.8	117.1	0.972	492	2210	2702	2771	2872	2972	3072	3175	3277
1.0	120.4	0.882	506	2201	2707	2770	2871	2971	3072	3174	3277
1.5	127.6	0.714	536	2181	2717	2766	2868	2970	3071	3173	3277
2.0	133.7	0.604	562	2163	2725	2761	2866	2968	3070	3172	3275
2.5	139.0	0.522	585	2148	2733	2757	2864	2767	3069	3171	3274
3.0	143.8	0.462	605	2133	2739	2752	2862	2965	3067	3170	3274
3.5	138.9	0.414	624	2120	2744	2748	2860	2963	3066	3169	3273
4.0	151.9	0.375	641	2108	2749	—	2856	2961	3064	3168	3272
5.0	165.1	0.315	671	2086	2757	—	2850	2958	3062	3166	3270
7.5	172.9	0.227	733	2039	2772	—	2837	2950	3056	3161	3267
10.0	184.1	0.177	782	2000	2782	—	2823	2940	3049	3156	3262
12.5	193.4	0.146	823	1966	2789	—	2806	2930	3043	3149	3258
15.0	201.4	0.124	859	1935	2794	—	—	2921	3036	3146	3254
17.5	208.6	0.108	891	1907	2798	—	—	2910	3029	3139	3251
20	215.2	0.095	920	1879	2799	—	—	2899	3022	3136	3246
25	226.0	0.080	972	1831	2803	—	—	2876	3005	3126	3237
30	235.5	0.066	1016	1787	2803	—	—	2857	2992	3115	3229
35	243.7	0.058	1058	1744	2802	—	—	2853	2982	3103	3224
40	251.7	0.050	1095	1706	2801	—	—	—	2959	3091	3212
50	265.0	0.039	1161	1632	2793	—	—	—	2923	3067	3194
60	276.6	0.032	1219	1564	2783	—	—	—	2882	3041	3175
70	286.3	0.028	1272	1499	2771	—	—	—	2831	3015	3156
80	295.8	0.023	1321	1435	2756	—	—	—	2770	2987	3137
90	303.8	0.021	1368	1373	2741	—	—	—	—	2956	3116
100	311.6	0.018	1412	1311	2723	—	—	—	—	2922	3095

Properties of Common Substances

Substance	*Density kg/m³*	*Specific Heat J/kg °C*
Alumina	3800	963 – 1130
Asbestos	2000 – 3000	837 – 1047
Ashes	608	879
Asphalt	1100 – 1400	921
Brickwork	1500 – 1800	837
Cellulose (dry)	1400	1340 – 1549
Cement	1300 – 3000	796 – 837
Chalk	1900 – 2800	796
Charcoal	500 – 1600	670 – 1005
Clay (dry)	1800 – 2500	921
Concrete	2000	670 – 921
Cork	250	2052
Firebrick	2000 – 2700	963 – 1089
Glass–silicate	2700	796 – 1089
Graphite	2250	837
Gypsum	2320	1089
Leather	860	1507
Limestone	2680 – 2760	921

Pressures and Barometer Readings at Different Altitudes

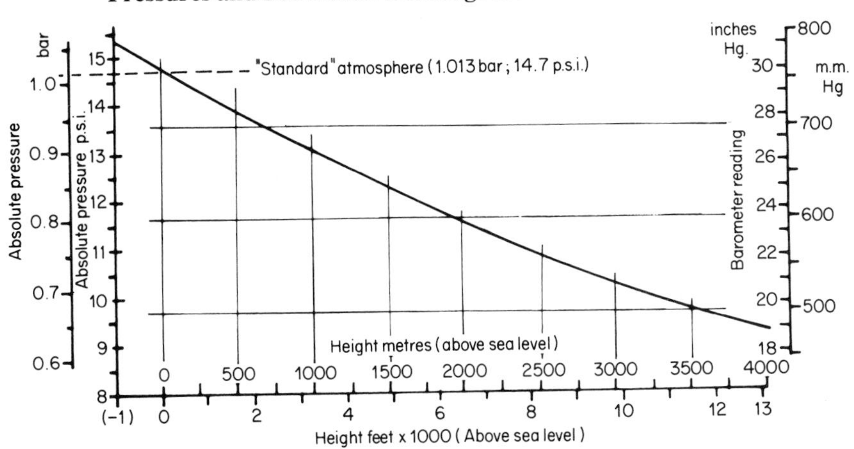

Properties of Refractories

	Specific Heat J/kg°C		Linear % Expansion 20-1000°C	Thermal Conductivity		Bulk Density kg/m³
	0-500°C	0-900°C		W/m³ 500°C	900°C	
40% alumina brick	980	1000	0.60	1.20	1.23	2300
95% alumina brick	1025	1120	0.80	3.20	2.55	2900
Fireclay brick	955	1070	0.90	1.25	1.35	2350
Silica brick	1005	1075	1.30	1.40	1.55	1790
Magnesite brick	1100	1175	1.20	6.20	4.50	2790
Chrome—magnesite	940	1030	0.90	1.70	1.65	2950
Chrome—brick	860	930	0.85	2.25	2.15	3100
Silicon carbide	720	780	0.60	13	12	2550
Insulating brick (1250°C grade)	900	1000	0.90	0.30	0.38	800
Zircon	690	750	0.50	2.50	2.38	3750
Castable insulating "concrete" (1300°C)	—	—	—	0.52	0.56	—
Ceramic fibre blanket (64kg/m³)	—	—	—	0.17	0.31	64
Ceramic fibre blanket (128 kg/m³)	—	—	—	0.12	0.20	128
Glass or mineral wool slab	—	—	—	0.09	—	160

The linear expansion quoted is the reversible change in length. Some refractories can also undergo a permanent change in dimensions at high temperatures.

Density of Water at Atmospheric Pressure

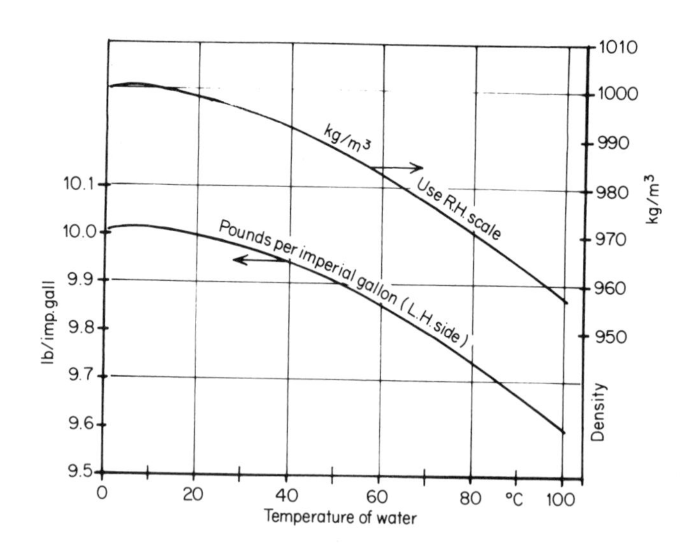

Properties of Various Steels

In Table 1 = high purity iron; 2= mild steel, 3 = medium carbon steel;
4 = nickel-chromium steel; 5 = nickel/chrome/molybdenum steel;
6 = "18/8" chromium/nickel steel; 7 = stainless steel

Material		*1*	*2*	*3*	*4*	*5*	*6*	*7*
Composition % by weight	C		0.23	0.43	0.32	0.34	0.08	0.27
	Si		0.11	0.20	0.25	0.27	0.68	0.20
(typical)	Mn		0.63	0.70	0.55	0.55	0.37	0.28
(balance is Fe.)	Cr		—	0.03	0.70	0.80	19.1	13.70
	Ni		trace	0.04	3.40	3.50	8.1	0.20
	W		—	—	—	—	0.6	0.25
	Mo		—	trace	trace	0.40	—	trace
Thermal linear Expansion (Coefficient × 10^6/°C)								
Over range 0–100°C		12.4	12.2	11.6	11.5	11.6	14.8	10.0
0–300°C		12.7	13.1	13.1	12.7	12.6	17.1	11.1
0–500°C		13.8	13.9	14.2	13.6	13.5	18.0	11.8
0–700°C		14.5	14.9	15.1	13.7	13.5	18.8	12.5
0–900°C		—	12.4	13.6	12.1	12.0	19.2	10.6
Specific Heat J/kg°C								
Over range 0–100°C		452	477	470	478	478	500	465
0–300°C		495	511	515	520	515	530	510
0–500°C		540	550	550	553	558	546	555
0–700°C		590	615	607	640	640	570	625
Thermal Conductivity (W/m °C)								
Over range 0–200°C		62	51	48	36	34	16	26

Representative Properties of Metals other than Steels (Compiled from several sources)

Metal	*Specific Gravity (Water = 1 at 15°C)*	*Coefficient of Expansion m/1,000,000 m/°C*	*Thermal Conductivity W/m °C*	*Specific Heat J/kg °C*
Aluminium	2.7	12.7	204	946
Antimony	6.6	6.3	19	205
Bismuth	9.8	7.5	8.1	121
Chromium	7.2	4.7	272	502
Cobalt	8.9	6.7	69	410
Pure Copper	8.9	9.3	381	385
Admiralty and 70% brass	8.5	10.3	100	—
Muntz and yellow brass	8.4	11.6	121	385
Red brass	8.6–8.75	9.8–10.4	159	—
Bronze	8.8	9.8	173	—
Everdur	8.2	7.8	28	—
Monel	8.8	7.8	26	532
Mercury (liquid above –39°C)	13.5	61	8 (liquid)	—
Gold	19.3	14.1	312	131
Lead	11.3	16.4	33	126
Magnesium	1.74	14.3	159	1042
Nickel	8.9	7.3	61	469
Platinum	21.4	9.0	73	136
Silver	10.5	10.5	415	234
Sodium	0.97	39.5	45	1235
Tantalum	16.6	3.6	61	151
Tin	7.3	11.2	59	226
Uranium	19.1	—	30	116
Vanadium	19.3	2.2	199	142
Zinc	7.1	18.3	111	373

Weight of Air at Different Altitudes and Temperatures

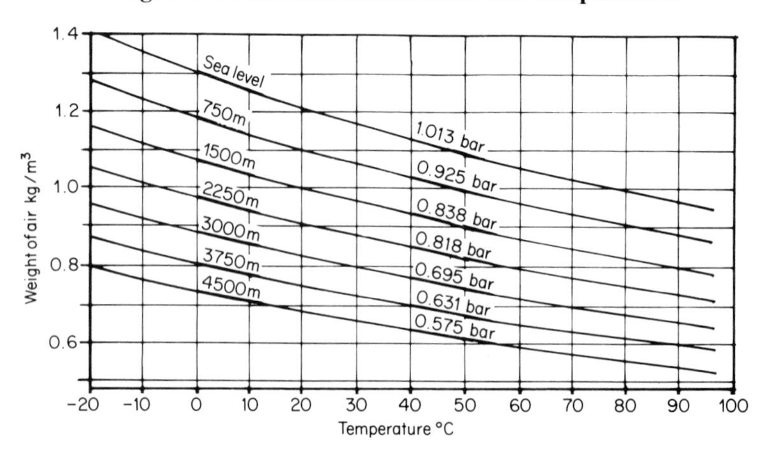

2. OFFICES, INDUSTRIAL BUILDINGS ETC. LEGISLATION ON INSULATION AND CONTROLS

Insulation of Structure

The Thermal Insulation (Industrial Buildings) Act 1957 set moderate standards for insulation of roofs of factories, ignoring walls and windows, applicable only where the building uses a dedicated space heating system. These requirements have now been overtaken by Building Regulations, the major statutory Instrument being No.1676 of 1976. This main document has been successively amended in 1978 (Building – First Amendment – Regulations 1978) No. 723 and 1981 (Building – Second Amendment – Regulations 1981) No. 1338. These apply to dwellings such as houses and flats, and offices, factories, storage buildings, shops, institutional buildings and places of assembly.

The following very brief summary can be given, although the above Regulations should be studied for full details. (Although strictly applicable to only England and Wales, corresponding legislation is being enacted for Scotland and Northern Ireland.)

External walls around heated spaces; internal walls exposed to a ventilated space, floors whose under-surface is exposed to external air or a ventilated space; roofs over heated spaces, including ceilings with a ventilated space above.

Maximum Average U Value in W/m² °C
For Factories & Storage Buildings: 0.7

For all other types of buildings such as shops, offices, institutional buildings & places of assembly: 0.6

Note: These are average *U* values – parts can be under-insulated if other parts receive extra insulation to compensate.

The windows in such buildings have maximum limits, related to percentage of wall or roof areas. If the full rooflight allowance is not used the balance can be re-allocated to wall windows and vice versa. For shops only the area of ground floor windows is not included and the area of ground floor walls is also not included in total wall area, which allows large display windows to be used without being counted in the permissible maximum areas for the rest of the building. Roofs with pitches of 70° or more are to be treated as external walls. Exemptions can be claimed where the space heating is only of low level or background type; less than 50 watts/m² of floor area for factories and storage, or 25 watts/m² for all other buildings. Small buildings under 30m² floor area are also exempt.

Maximum Permitted % of Total Area	*Factories and Storage*	*Offices, Shops, places of assembly*	*Institutional or residential*
Single glazed windows in walls	15	35	25
Single glazed rooflights in roofs	20	20	20

These percentages can be increased using double or triple glazing providing it is shown that the total heat loss would not exceed that from single glazed units complying with these maximum percentages.

Houses and flats have more stringent requirements for structural insulation and

2. OFFICES, INDUSTRIAL BUILDINGS ETC.

maximum permissible *U* Values in W/m^2 °C are:

Roofs 0.35; Walls 0.60; exposed floors 0.60

It must be pointed out that walls of normal brick/cavity/aerated block/plaster construction cannot achieve these 0.60 or 0.70 figures – extra insulation such as cavity fill or slab is necessary, or additional insulation added to the internal face of the blockwork. Window openings shall not exceed 12% of the perimeter wall area (24% for double glazing, 36% for treble glazing) although if the walls are insulated to a higher standard than 0.60 the window areas can be increased to use up the reduction in wall heat loss.

Pipes and Hot Water Storage

These same Building Regulations now stipulate, from April 1982, the minimum standards of insulation for pipes, ducts and storage vessels which carry heated gases or fluids in connection with the heating system of any building. Exemptions apply to items designed to contribute to space heating requirements, flexible connections, valve handles and control equipment and very short runs to hot water taps or outlets (for example up to 8m of pipe above 22mm and up to 28mm outside diameter can be used to supply a tap without the need for insulation), and these Regulations do not apply to industrial process items. For pipes coming under these Regulations, very briefly:

(a) Pipes with o.d. below 50mm should have an insulation thickness at least equal to the o.d.

(b) Pipes with o.d. of 50mm or more should have at least 50mm of insulation applied.

(c) Ducts and vessels should have enough insulation to reduce losses to a maximum of 90 W/m^2 (more detailed recommended thickness are in BS 5422:1977).

Remember these are minimum requirements – it may be economic to apply greater thickness.

Time Clocks, Thermostats and Controls

Any new heating system serving a floor area of more than 125m^2, or any new extension of an existing system serving a new floor area of more than 125m^2, must be fitted with thermostat control. This must be inside the building for each independently controlled system and outside the building where a central hot water heating system serves radiators or natural convectors. Where the heating is intermittent an automatic control must be fitted so that normal internal temperatures are only maintained when the building, or part of it, is occupied. Where the output rating of the new system or extension is 100kW or less a time clock will suffice.

Where the 125m^2 rule applies and there are two or more interconnected oil or gas fired boilers whose combined rating exceeds 100kW, automatic controls must be fitted to shut down and start up the boilers so that only the number of boilers are in use that are necessary to meet the building load and the flow of water is reduced through any boiler that is shut down.

Where there is a hot water storage vessel a thermostat is to be fitted and if the storage exceeds 150 litres an automatic switch is to be fitted to stop and start the heat supply at pre-set times.

2. OFFICES, INDUSTRIAL BUILDINGS ETC.

Likely Coefficients of Overall Heat Transfer for Typical Building Constructions (allowing for external and internal surface resistances and any air-filled cavities)

Roofs	*U Value W/m² °C*
Pitched, covered with slates or tiles, roofing felt underlay, plasterboard ceiling	1.70 – 2.0
Pitched, covered with slates or tiles, with fibre roll insulation above ceiling	
60mm	0.50
100mm	0.35
Pitched, covered with slates or tiles, with rigid insulating board replacing	
pasterboard 40mm	0.60 – 0.67
75mm	0.39 – 0.42
Pitched, covered with slates or tiles, rigid insulating lining board, 15 – 20mm fibre type	0.9
Steel or asbestos cement roofing sheets, no lining	6.1 – 6.7
Steel or asbestos cement, rigid insulating lining board, 15 – 20mm fibre	0.9
Steel or asbestos cement, rigid insulating lining board 50mm	0.60
Steel or asbestos cement, rigid insulating lining board 75mm	0.40
Pitched, covered with slates or tiles or asbestos cement sheeting, with flat ceiling of plasterboards, no felt	3.20
Pitched, covered with slates or tiles with fibre roll insulation laid above	
ceiling 60mm	0.50
80mm	0.40
Flat roof, 3 layers felt on chipboard or timber sheeting plasterboard ceiling	1.54
Flat roof 3 layers felt on rigid insulating board under felt roofing	
40mm	0.55 – 0.60
70mm	0.37 – 0.40
100mm	0.28 – 0.30
Flat roof 3 layers felt, glass fibre roll insulation laid above ceiling	
60mm	0.50
100mm	0.35
Flat roof 3 layers felt etc., on metal decking, no ceiling	3.3
Flat roof 3 layers felt etc., rigid insulating board between felt and deck	
40mm	0.70
70mm	0.50
100mm	0.3
Flat roof, felted, on concrete structured deck, plaster finish to underside	2.60
Flat roof, felted, on concrete structured deck, rigid lining board between	
concrete and felt 40mm	0.65
70mm	0.40
100mm	0.30

Walls	*U Value W/m² °C*
Steel or asbestos cement cladding sheets	5.3 – 5.7
Steel or asbestos cement cladding sheets with plasterboard lining and layer	
of fibre roll insulation 60mm	0.55
100mm	0.40
Steel or asbestos cement cladding sheets with rigid insulating board,	
decoratively faced 50mm	0.60
75mm	0.40
Brick/cavity/brick 11″ wall, internally plastered	1.4 – 1.9
Brick/Cavity/brick 11″ wall, cavity filled with loose, foamed, or slab insulation	0.55 – 0.60
Brick/cavity/brick 11″ wall, normal cavity, internal surface battened, fibre roll insulation, plasterboard cover 25mm	0.70
Brick/cavity/brick 11″ wall normal cavity, internal surface battened, fibre roll insulation, plasterboard cover 50mm	0.50 – 0.60
Brick/cavity/low density block 11″ normal wall, internally plastered	0.9 – 1.5

2. OFFICES, INDUSTRIAL BUILDINGS ETC.

Roofs	*U Value W/m² °C*
Brick/cavity/low density block 11″ normal wall, cavity filled with loose, foamed or slab insulation	0.45 – 0.60
Brick/cavity/low density block 11″ normal wall, normal cavity internal surface battened, fibre roll insulation, plasterboard cover 25mm	0.60
50mm	0.45 – 0.55
Floors (Where space below is ventilated or open to air)	
Timber floor (boards or chipboards) underside plasterboard or asbestos sheeting	1.50
Timber floor (boards or chipboards) with fibre roll insulation between joists	
60mm	0.50
80mm	0.40
Concrete slab with surface tiles or boards, underside plastered or sheeted	2.00
Concrete slab with insulation slabs, new screed to carry tiles or boards	
50mm	0.60
Glazing (assuming fairly large panes and allowing for effect of frames. Small panes in single metal frames can have appreciably higher heat losses due to frames acting as a heat path)	
Single glazing	5.7
Double glazing (sealed units)	4.3
Treble glazing (sealed double unit and separate inner sheet)	3.2

The values shown are for "normal" weather conditions and site exposures. They may increase appreciably for severe weather or exposures.

These values also vary slightly with material from different manufacturers. Generally the lower range of insulating values (i.e. higher "U" values) are quoted. Where "rigid board" is mentioned, this is of glass or mineral wool composition. Slightly lower "U" values are possible if rigid polyurethane or polyisocyanate boards are substituted, but these can create dense smoke problems in fire situations.

Degree Days

Heating degree days can be defined as the mean number of degrees by which the outdoor temperature on a given day is less than the base temperature, added up for all the days in the period.

Cooling degree days are similarly defined as the mean number of degrees by which the outdoor temperature on a given day is above the base temperature, added up for all the days in the period.

In Britain, the base temperature for heating degree days is 15.5°C (60°F approx.). Tables of monthly degree days are published in the Department of Energy's monthly newspaper *Energy Management* and their *Fuel Efficiency Bulletin* No.7 discusses their use in comparing fuel usage on heating systems year by year or monthly. Remember the normal published Tables are for 24 hour days so really are only fully relevant to three-shift occupancy of buildings. Older tables were in degrees Fahrenheit! More recently the *Energy Management* Tables are in °C. Make certain you know which type of degree is used in any Table you find in textbook or journal!

Degree days (15.5°C basis) can be converted to Degree days (60°F basis) by multiplying the figures by 1.8.

2. OFFICES, INDUSTRIAL BUILDINGS ETC.

The two diagrams following show, first, the isopleths (contours of equal value) over Britain so that the degree-days per year can be seen for typical locations. The Tables that are published are based on 17 reporting stations and care should be taken not to assume these apply to the whole region in the Table. It will be seen from the Diagram, for example, that Heathrow, Glasgow and Manchester, 3 of these reporting stations, happen to lie about exactly on a "contour" so that only a few miles from these stations the Degree day figures can differ considerably.

The second diagram uses this information to give a rather simplified presentation, showing average Winter temperatures over Britain for the heating season period. Remember these are averages, October to April, and do not necessarily relate to the minimum Winter temperature. It can be seen that with absolutely identical buildings, one in Leeds and the other in Bournemouth, the Leeds plant would use more fuel to give identical hours of heating because the average Winter temperature is 6°C compared to 8°C in Bournemouth. (Referring to the first diagram showing degree day totals, Leeds has 2,800 (on the "contour" between two areas) and Bournemouth around 1,940–2,200.)

Reminder The 1980 Fuel and Electricity (Heating) (Control) (Amendment) Order sets a legal maximum temperature at 19°C (66.2°F) for heated buildings other than domestic premises! The Factories Act 1961 sets a minimum temperature of 15.5°C (60°F) and the Shops, Offices and Railway Premises Act 1963 sets a minimum of 16°C (60.8°F).

Degree day tables are based normally on 15.5°C. This does not mean that internal space heating temperatures should be as low as this, but reflects the situation that other heat gains from lights electrical equipment and people also contribute to keeping a building warm so that the "official" heating system can usually be turned off when outside temperatures reach 15.5°C.

Town Key The two diagrams overleaf show the position of various towns but to avoid too much clutter these are referred to by one or two letters only: The key is:

A = Aberdeen
B = Bristol
C = Cardiff
CL = Carlisle
G = Glasgow
L = London
LV = Liverpool
N = Norwich
S = Sheffield
BE = Belfast
BN = Bournemouth
CK = Cork
E = Edinburgh
GT = Gatwick
LD = Leeds
M = Manchester
NC = Newcastle
H = Hull

2. OFFICES, INDUSTRIAL BUILDINGS ETC.

Degree Day Isopleths

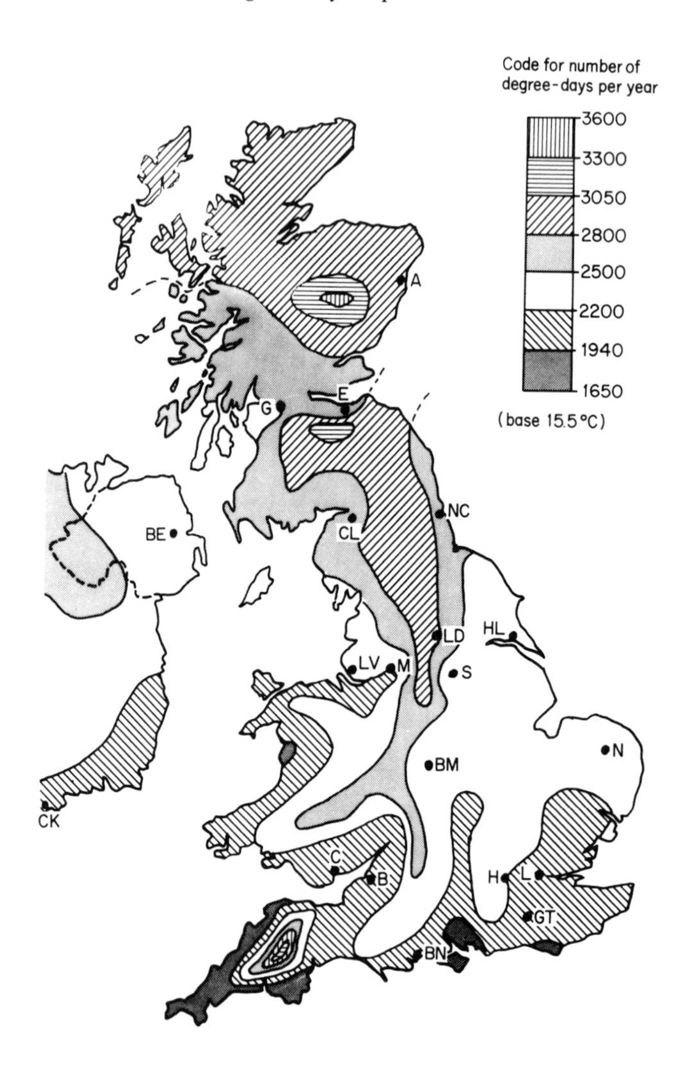

Average Temperatures During Heating Season

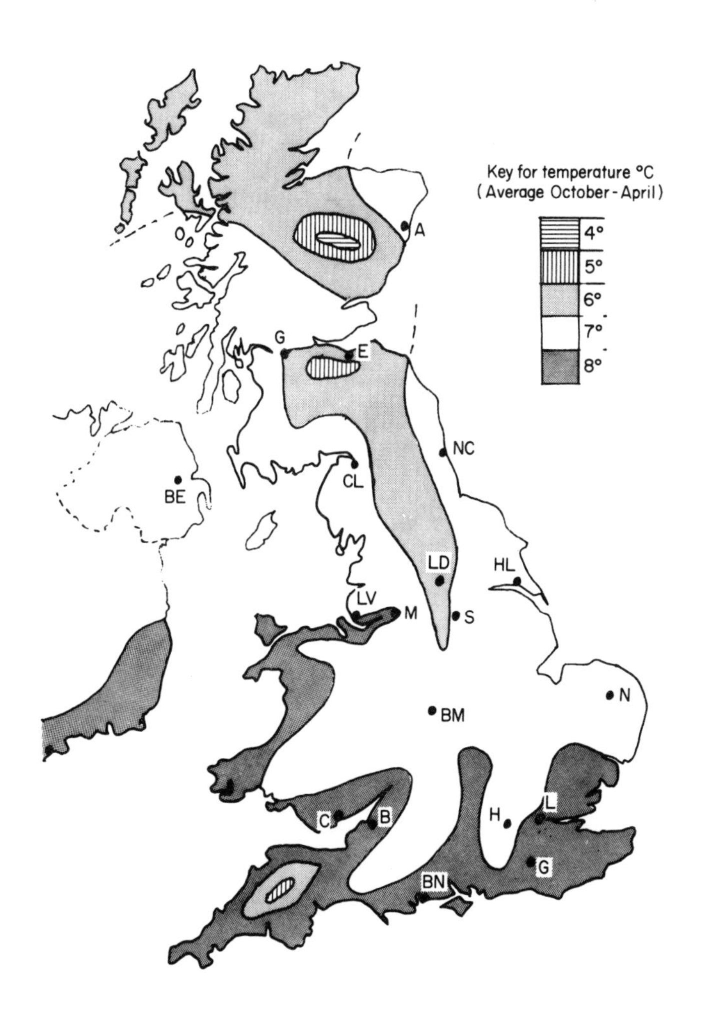

3. FUELS

Oil

Oil fuels are distillate or residual oil blends produced from the refining of crude oil and manufactured to a standard given in BS2869:1983, *Fuel oils for oil engines and burners for non-marine use*.

The Standard defines 9 classes and subclasses of oil product and sets limits for their physical properties such as viscosity, cetane number, flash point, water, ash and sulphur content, etc. It does not specify calorific value. The principal property on which fuels are classified is viscosity index and they are usually known by their generic names, class designation or the viscosity index on the Redwood scale. (This is an older method of determining viscosity but reference to Redwood viscosities of oils is still common. On the Continent it is not uncommon to refer to oils by their Saybolt equivalents.) These are given below.

Oil suppliers publish a schedule of prices for their products according to their own delivery patterns. This is usually based on delivery zones. For most suppliers the pricing zones are the same and the price differences between zones are the same. The differential increases at present are 0.04 pence per litre from inner to outer, 0.04 pence per litre from outer to general. The oil companies have not published precise zone boundaries for some time.

Full load volumes and small load surcharges vary by company. In addition to these there are various discounts offered so that the actual price paid may be less than the scheduled price.

Standard Grades and Properties of Oil Fuels

Product generic name	Gas oil			Light Fuel Oil			Medium Fuel Oil			Heavy Fuel Oil		
BS 2869 class description	D			E			F			G		
Approx. Redwood viscosity index (seconds)	35			200			950			3500		
Typical UK Specifications	*min.*	*max.*	*typical*	*min.*	*max.*	*typical*	*min.*	*max.*	*typical*	*min.*	*max.*	*typical*
Kinematic viscosity at 40 °C (cSt)	2.0	5.0	3.2									
Kinematic viscosity at 80 °C (cSt)				10	13.5	13	25	35	34	55	85	78
Density at 15 °C (kg/litre)	0.820		0.845		0.980	0.940		0.990	0.970		1.005	0.980
Other UK data:												
Specific volume (litre/tonne)		1220	1183	1042		1064	1010		1030	1005		1020
Calculated calorific value, gross (GJ/te)	45.2		45.5	42.5		43.3	41.8		42.7	41.6		42.4
Calculated calorific value, net (GJ/te)	42.4		42.7	40.1		40.8	39.5		40.3	39.3		40.4

Oil companies have recently announced increases in maximum viscosities of fuel oils: E to 220; F to 1000; G to 4000/4500, all Redwood viscosities. These higher viscosities will require slight increases in preheat temperatures.

Fuel Oils (Brackets show approx. viscosity as Redwood No.1 seconds at 100°F)	*Light ("200")*	*Medium ("950")*	*Heavy ("3500")*
Typical temperatures °C — handling from storage	10	30	45 to 50
— in storage	10	25	35 to 40
— pumping	10	25 to 30	45 to 55
— at burners			
(a) Pressure Jet	60–75	95–105	120–130
(b) air or steam atomised	45–70	75–110	100–125
(c) rotary cup	35–45	60–70	80–90
Typical sulphur content %	2.5–3.2	3.5	3.5
Specific heat relative to water	0.46	0.45	0.45

Note: In recent years the storage, pumping and atomising characteristics of medium and heavy fuel oils have been slowly altering due to changes in sources of crude oil, and in refinery practices. It is recommended that the upper temperatures in the above Table should be used, and in case of doubt the supplier should be consulted.

Viscosity of Fuel Oils at Various Temperatures

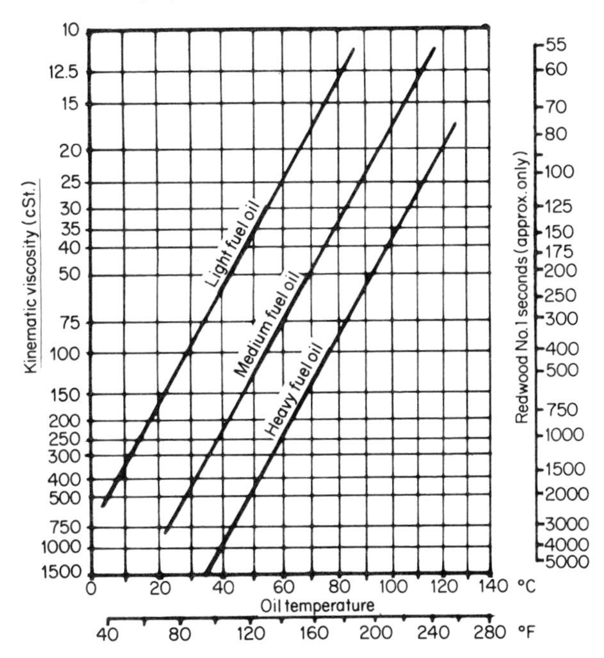

Fuel Oil Pricing Zones

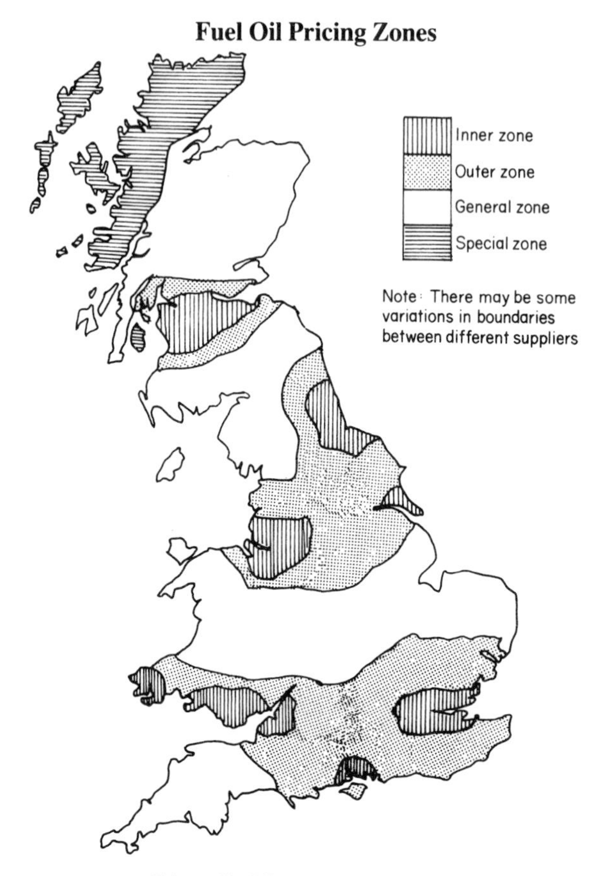

Liquefied Petroleum Gases

Liquefied petroleum gases are hydrocarbon gases which can be liquefied under pressure. There are two: propane and butane. The requirements for the quality of commercial propane and butane are set out in British Standard BS4250: 1975, *Specifications for Commercial Propane and Commercial Butane*.

Although they are called propane and butane they are in fact mixtures of hydrocarbons, commercial propane consisting of mainly C_3 hydrocarbons (that is hydrocarbons with 3 carbon atoms in the molecule) and commercial butane consisting of C_4 hydrocarbons so they may contain substantial and variable proportions of propylene and butylene. They are each allowed small quantities of lighter and heavier hydrocarbons. Since the calorific values of other constituents differ from those of pure propane and butane there can be small variations in calorific value also but for the limits set for other constituents this variation is likely to be within one or two per cent and affects propane more than butane. Typical properties of commercial propane and butane are given below. Commercial butane is sold by weight. Commercial propane is either sold by weight or volume.

Typical properties of commercial propane and butane

	Commercial butane	*Commercial propane*
Litre/tonne at 15°C	1723–1760	1965–2019
Relative density of gas compared with air at 15°C and 1016 mbar	1.90–2.10	1.40–1.55
Volume of gas (litres) per kg of liquid at 15°C and 1016 mbar	406–431	537–543
Ratio of gas volume to liquid volume at 15°C and 1016 mbar	233	274
Boiling point at atmospheric pressure °C approx	–2	–45
Latent heat of vaporisation (kJ/kg) at 15°C	372	358
Specific heat of liquid at 15°C (kJ/kg deg C)	2.386	2.512
Calorific values:		
Gross (MJ/kg)	49.3	50.0
Net (MJ/kg)	45.8	46.3
Sulphur Content (%)	Negligible to 0.02%	Negligible to 0.02%

Coal

Coal at the mine is analysed by continuous sampling for calorific value, ash, moisture and sulphur content. It is sold by weight but its price in £s per tonne is determined from a points scale based mainly on gross calorific value with other adjustments added. How this operates is described in a recent CBI Note (B272 83, 1 November 1983). The basic points are obtained by dividing the as received calorific value in kJ/kg by 100, multiplying this by an area points value, which depends on the NCB Area, to obtain the Unadjusted Notional Price. An adjustment is then made by subtracting points for Secondary Ash Content, adding points for Grade and adding or subtracting points for the Sulphur Allowance. Finally, the points value is rounded by a formula which rounds some prices up and others down, a process called Boxing. The final answer is the price in £ per tonne. On top of that then has to be added delivery costs.

Delivery prices can be somewhat variable. The basic price of coal does not depend on the quantity you buy but this will affect delivery costs in some cases.

Gross calorific values of coal sold to industry can vary between 24 GJ/tonne and 31 GJ/tonne depending on type and origin.

Grade is a term used only in relation to the physical size of coal, a graded coal being one which has a limitation upon both its maximum and minimum particle sizes. As well as graded coals such as singles (prepared through screens not less than 25mm but not greater than 38mm and over screens not greater than 18mm but not less than 12.5mm) there are others which have only a single size limitation; smalls have a specified upper size limit, usually below 50mm but no minimum size; fines are small coal with a top size usually below 3mm; large coals have an agreed minimum size but no upper limit. An untreated coal is one which has not been subjected to cleaning processes, but industrial coals are normally washed, which leaves them free of loose dirt.

The National Coal Board does not publish much by way of detailed general specification of solid fuel. There is, for example, no recent summary available of the properties of coals which the Board sells. However, this is largely only due to the amount of information for the large number of coal sources available and the problems of summarising it in a way that would be useful. Enquiries regarding coal characteristics made directly to the Board's Technical Service will usually get a prompt response.

Typical Properties of Some Industrial Coals

NCB Rank Code Number		100	600	700	800
Typical Analysis as fired: (% by weight) (analyses are for marked coals)	Moisture	6	11	11–12	12
	Carbon	80	68	65–66	61–62
	Hydrogen	2–3	4.5	4	4
	N_2 and O_2	2–3	7	7–8	8–9
	Sulphur	1	1.7	1.7	1.7
	Ash	8	8	8	8
Typical volatile matter (dry ash free coal) %		5–6	30	30–31	31
Typical calorific value as fired, Gross		30.9	28.4	27.5	26.7
(MJ/kg) calorific value as fired. Net		30.0	27.2	26.2	25.4

100 = anthracite, non-caking, for small boilers, cookers, etc.
600 = medium caking, used for open fires, for coke-ovens (often blended with stronger-caking coals)
700 = weakly caking, mainly used for industrial boilers
800 = very weakly caking, smaller industrial boilers

Natural Gas

Natural gas is mainly methane which occurs naturally underground trapped in rock strata. Most of the natural gas in the UK comes from beneath the North Sea. Gas is metered by volume since gas from the different fields varies slightly in composition, it is then corrected for calorific value.

The broad policy on the marketing of gas is to recognise its premium qualities of clean burning and ease of handling and to give preference to premium applications and to make it available to non-premium users on an interruptible basis mainly to assist in system balancing in the supply network. This is important because as a fuel "on tap" gas is affected by short-term and seasonal variations in demand.

Gas is metered by volume and is regularly tested for calorific value per unit standard volume. For Published Tariffs there is a general declared calorific value and for contract customers the actual calorific value is applied.

There are three tariffs. First is Published Tariff Gas available to users of less than 25,000 therms (2500GJ) a year, and so is appropriate only to small users. The tariff for this is published. Users over 25,000 therms (2500GJ) a year are offered Special Contract terms. Such contracts are confidential between the Gas Regions and their customers and fall into two categories – firm and interruptible. Firm gas is a continuously available supply generally only available for premium uses. Interruptible supply is one in which the Gas Region can cut off supply to the customer when management of the supply network requires it. To take interruptible gas, the customer must be prepared to contract for typically 63 days interruption a year. Contracts used to be related closely to oil products – gas oil for firm supplies and heavy fuel oil for interruptible supplies. However, gas prices have been pegged for sometime and the price advantage of gas against fuel oil products has widened over the last three years.

Typical Properties of Natural Gas

Composition (% by volume): (S is negligible)	CH_4	94.4
	C_2H_6	3.0
	Other Hydrocarbons	0.9
	CO_2	0.2
	N_2	1.5
Density (kg/m^3 at 15°C, 1.013 bar)		0.73
Density relative to air		0.59
Calorific values (MJ/m^3 at 15°C, 1.013 bar)	38.62 gross	34.82 nett

Electricity

The bulk of our electricity in the UK is generated in large centralised coal, oil and nuclear power stations and distributed through the public supply.

The electricity supply industry is a highly capital intensive industry supplying a commodity essentially on instant demand. In setting its tariffs, the industry takes into account the differences between consumers in terms of capital deployed to connect the consumer through the supply network, the pattern of demand and the cost of generation.

Electricity, as such, cannot easily be stored, and the supply industry has to reconcile this with the fact that demand is exceedingly variable. Large simultaneous demands on the grid create great difficulties for the supply industry and if it had no means to manage demand it would require a very much greater surplus of expensive generating plant. The industry therefore needs to manage demand and does this through the tariff structure. Mainly this is to make off-peak electricity cheaper and on-peak more expensive.

Marketing of electricity is the responsibility of the Area Boards. Each Area Board sets and publishes its own tariffs and the tariff structure can vary from one Board to another. From time to time the tariffs available can vary.

Tariffs are designed to encourage consumers to manage their load pattern. They usually, but not always, contain a standing charge element and a variable cost element related to consumption. The variable cost part often includes a block of units at one price and further units at reduced price. There are special rates applied to supplies which are restricted to certain times of day. Many industrial consumers operate on maximum demand tariffs into which off-peak features can be incorporated.

It is frequently the case that consumers have open to them a number of tariff structures and can select a tariff which is of advantage to them. Advice on tariff selection is usually available from your local Area Board Office.

1 Unit of electricity (kWh) has a heat equivalent of 3414 Btu, or 3.60 MJ, if converted to heat at 100% efficiency.

3. FUELS

Current Production of North Sea Oil Fields

Figures are in units of 1000 US barrels per day (1 barrel = 35 imperial gallons or equivalent to approximately 0.23 tonnes of coal). The figures shown in brackets at the side of the barrel outputs indicate the equivalent coal quantity in 1000 tonnes units. Altogether, some 31 fields are currently producing or under active development. The 25 listed below have a combined production around 2.5 million barrels per day – around 365,000 tonnes of oil per day or 575,000 tonnes coal equivalent. These 31 fields are estimated to have reserves of around 1100 million tonnes of oil (about 8 years at the present extraction figures) but the total reserves are probably three times this figure.

Brent	470 (108)	Magnus	90 (21)	N.W. Hutton	46 (11)
Forties	420 (97)	Maureen	90 (21)	Beatree A	36 (7)
Ninican	280 (65)	Beryl A	80 (18)	Buchan	30 (6)
Piper	195 (45)	Dunlin	74 (17)	Heather	26 (6)
Fulmar	115 (26)	Brae	65 (15)	Duncan	25 (6)
Thistle	105 (24)	N. Cormorant	47 (11)	Tartan	25 (6)
Claymore	100 (23)	C. Cormorant	40 (9)	Argyll	21 (5)
Murchison	95 (22)	S. Cormorant	40 (9)	Montrose	14 (3)
				Auk	12 (3)

Average Cost of Fuels to British Industry

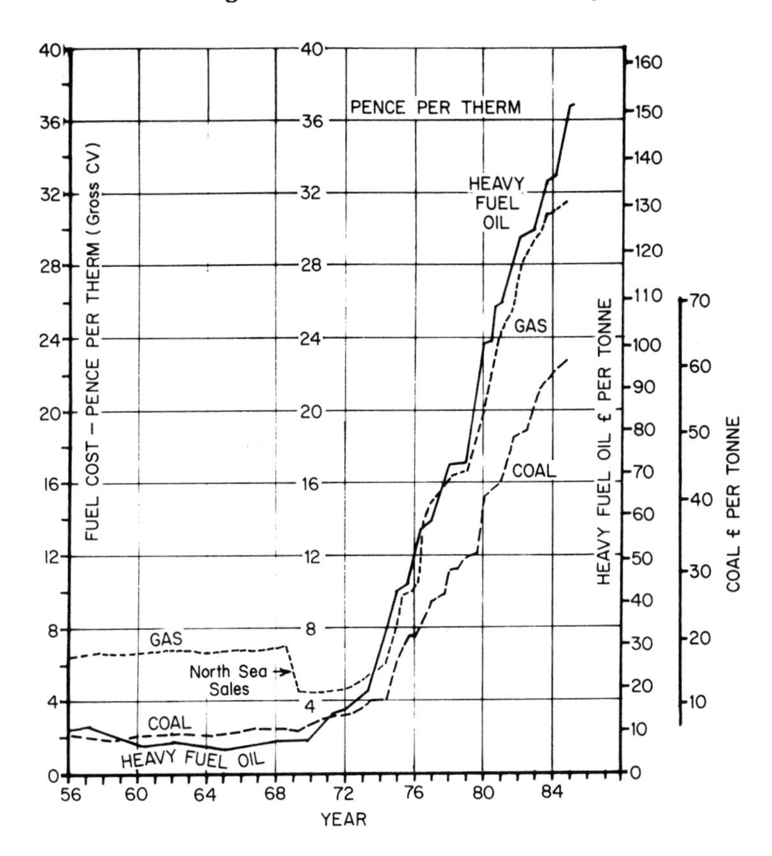

This diagram shows that fuel prices remained almost constant for 14 to 15 years up to 1970. Only gas was affected near the end of this period as natural gas, brought ashore from the North Sea, was sold in cheap bulk contracts. After 1970 the OPEC price increases caused massive rises in oil prices which caused gas and oil to follow due to market and inflationary pressures. Propane and butane (LPG) sell at approximately 11p/therm above the price of heavy fuel oil (about 48 – 50p/therm in late 1984).

Electricity cannot be shown on this diagram as its cost in terms of heat equivalent is completely off the scale. Typical figures in p/therm are :

1974 – 26 1978 – 65 1982 – 96
1976 – 52 1980 – 79 1984 – 109

(These are direct supply prices – off peak electricity prices are about 47 – 50% of these figures.)

Fuel Cost Conversion Table

	To / From	p/therm		p/kWh		£/GJ	
		Gross	*Net*	*Gross*	*Net*	*Gross*	*Net*
Fuel Type	*Cost Unit*						
Electricity	p/kWh	29.307	29.307	1.0	1.0	2.778	2.778
Natural Gas	p/cu.ft	96.617	107.113	3.297	3.655	9.158	10.152
Natural Gas	p/therm	1.0	1.108	0.0341	0.0378	0.0948	0.105
35sec Fuel Oil	p/litre	2.769	2.955	0.0945	0.101	0.262	0.280
200sec Fuel Oil	p/litre	2.612	2.762	0.0891	0.0942	0.248	0.262
950sec Fuel Oil	p/litre	2.599	2.748	0.0887	0.0938	0.246	0.260
3500sec Fuel Oil	p/litre	2.567	2.719	0.0876	0.0928	0.243	0.258
Propane	£/tonne	0.211	0.228	0.0072	0.0078	0.020	0.0216
Butane	£/tonne	0.214	0.230	0.0073	0.0079	0.0203	0.0218
Coal – Singles	£/tonne	0.372	0.388	0.0127	0.0132	0.0352	0.0368
Coal – Smalls	£/tonne	0.375	0.392	0.0128	0.0134	0.0356	0.0372
Industrial Coke	£/tonne	0.378	0.384	0.0129	0.0131	0.0358	0.0364

4. PRINCIPLES OF COMBUSTION

Combustion is the rapid chemical combination of oxygen with the combustible elements of a fuel. The most important fuels used commercially in industry are coal, residual fuel oil, distillate oils, LPG, natural gas (methane) and carbon monoxide/hydrogen gas (dying out). There are two combustible elements of principal significance in commercially used fuels, namely carbon and hydrogen. Sulphur and Chlorine are also present in some fuels and are of minor significance as a source of heat but of major concern in terms of corrosion and environmental problems.

Carbon and hydrogen when completely burned with oxygen yield carbon dioxide and water respectively as the combustion products :

$C + O_2 \rightarrow CO_2$

$2H_2 + O_2 \rightarrow 2H_2O$

The carbon to hydrogen ratio varies significantly from fuel to fuel. For example, anthracite is nearly pure carbon while natural gas (methane) contains 4 atoms of hydrogen for each atom of carbon. So for any fuel of composition $C_x H_y$:

$$C_x H_y + \left(x + \frac{y}{4}\right) O_2 = x\,CO_2 + \frac{y}{2} H_2O$$

The amount of oxygen theoretically required to burn a fixed quantity of fuel depends on the values of x and y, or in other words the carbon to hydrogen ratio of the fuel. The quantity of oxygen theoretically required is referred to as the *stoichiometric requirement* for complete combustion. This provides an important reference point against which the actual combustion conditions can be compared.

In practice, air is used for combustion rather than pure oxygen and data and tables for combustion efficiency calculations are usually related to air. In addition to the stoichiometric air requirement, it is also possible to calculate the theoretical yield of carbon dioxide assuming stoichiometric combustion. This calculation assumes that all carbon in the fuel is converted to carbon dioxide. The *theoretical carbon dioxide yield* is usually expressed as a *volume percent of dry flue gas*. The dry flue gas basis removes uncertainties caused by the moisture content of combustion air and makes practical carbon dioxide measurements easier. Stoichiometric air requirements and carbon dioxide yields for a range of fuels are shown in Tables.

Air

Typical composition of air is approx. :

	O_2		N_2
By weight	23.2%		76.8%
By volume	21%		79%
By molecules	1	to	3.76

More precisely, dry air by volume is 78.09% N_2, 20.95% O_2, 0.93% argon and other inert gases and 0.03% CO_2. Air used for combustion is rarely dry and usually contains water vapour, the amount depending on relative humidity and temperature. Examples at 25 °C (298.15K) are :

20% relative humidity = 0.004 kg H_2O/kg dry air

90% relative humidity = 0.020 kg H_2O/kg dry air

This water vapour passes through the boiler or furnace and carries heat away to waste, the amount depending on whether it remains as vapour (retaining its latent heat) or is condensed (containing sensible heat only). The latent heat of water at atmospheric pressure is 40.66 kJ/mol or 2257kJ/kg. In temperate and cold climates the loss of heat due to the water vapour in air is so small in comparison to other heat losses that it is usually ignored in calculations.

"Standard" conditions for quoting volumes, densities etc.

Older readers may think of NTP (normal temperature and pressure) conditions of 0 °C (32 °F) and 760mm Hg (101.325 kPa or 1.013 bar) or of STP of 60 °F (15.5 °C) and 30"Hg (762mmHg or 1.016 bar). However, in SI Units the normal Reference Temperature and Pressure is 298.15K (25 °C or 77 °F) and 101.325 kPa (1.013 bar). This will be used in the Tables etc. in this Combustion Section.

Calorific Values

Fuels containing hydrogen produce water vapour during combustion. The lower or net calorific value is that obtained if all the products of combustion, including water vapour, are uncondensed even though fully cooled back to the ambient temperature.

The higher, or gross, calorific value assumes that any water vapour is fully condensed so that its latent heat is recovered. Most boilers and furnaces are unable to condense this vapour and so many people feel boiler efficiences for comparative purposes should be calculated using the net calorific value of the fuel. Recent developments in central heating boiler design using extended heating surfaces and low return water temperatures have produced units which can condense at least part of the water vapour and so recover some additional heat. However, if fuels contain sulphur or chlorine such condensate can be very acid and cause corrosion and disposal problems. This really limits such boilers to the use of natural gas or LPG fuels which have negligible sulphur contents.

Thermodynamic Properties of Some Solids and Gases Concerned in Combustion Processes

All figures are in kJ/g.mole at standard conditions of 298.15K and 101.325kPa except for last column. Enthalpies of combustion refer to the process $C_x H_y + (x + {}^y/_4) O_2 = xCO_2 + {}^y/_2 H_2O$ (liquid)

Substance	*Enthalpy of Formation H_f*	*Combustion reaction*	*Enthalpy of Combustion $-\Delta H_c^\circ$*	*Entropy of Reaction ΔS°*	*Gibbs Function of Formation ΔG°*	*Weight of one mole in gms*
C (graphite)	–	$C + O_2 = CO_2$	393.51	+ 2.91	−394.38	12
C (graphite)	–	$C + ½ O_2 = CO$	110.53	+ 89.31	−137.16	12
C (graphite)	–	$C + CO_2 = 2 CO$	172.45	+175.71	+120.06	12
CO	−110.53	$CO + ½ O_2 = CO_2$	282.98	− 86.4	−257.22	28
H_2	–	$H_2 + ½ O_2 = H_2O$ (liquid)	285.83	−163.14	−237.19	2
H_2	–	$H_2 + ½ O_2 = H_2O$ (gas)	241.82	− 44.38	−228.58	2
H_2S	− 20.17	$H_2S + 1½ O_2 = H_2O + SO_2$	562.46		− 91.29	34
S	–	$S + O_2 = SO_2$	297.43		−152.01	32
CH_4	− 74.46	$CH_4 + 2O_2 = CO_2 + 2H_2O$	890.07	− 80.62	− 50.04	16
C_2H_6	− 83.82	$C_2H_6 + 3½ O_2 = 2CO_2 + 3H_2O$	1560.07	−174.08	− 31.09	30
C_3H_8	−104.68	$C_3H_8 + 5O_2 = 3CO_2 + 4H_2O$	2219.02	−269.59	− 24.03	44
C_4H_{10}	−125.65	$C_4H_{10} + 6½ O_2 = 4CO_2 + 5H_2O$	2877.05	−365.72	− 16.61	58

Notes: $\Delta G^\circ = \Delta H - T/ \Delta S = RT \ln K$
Commercial "propane" is over 90% $C_3 H_8$
Commercial "butane" is nearly 90% $C_4 H_{10}$
The weights of one gm. mole are "rounded" to nearest unit.

Properties of Gases Concerned in Combustion
(all volumes, densities and calorific values are at 298.15K (25°C) and 101.325kPa (1.013 bar)

Gas		*Molecular weight*	*Gas density kg/m³*	*Calorific value MJ/m³*		*Stoichiometric combustion requirements vol/1 vol fuel*		*Stoichiometric combustion products when burned in air vol/1 vol fuel*				
				gross	*nett*	O_2	*air*	CO_2	H_2O	N_2	*total (dry)*	*total (wet)*
Oxygen	O_2	32.00	1.308	–	–	–	–	–	–	–	–	–
Pure nitrogen	N_2	28.01	1.145	–	–	–	–	–	–	–	–	–
Air	–	"28.96"	1.184	–	–	–	–	–	–	–	–	–
Carbon dioxide	CO_2	44.01	1.799	–	–	–	–	–	–	–	–	–
Carbon monoxide	CO	28.01	1.145	11.57	11.57	0.50	2.38	1.00	–	1.88	2.88	2.88
Sulphur Dioxide	SO_2	64.06	1.107	–	–	–	–	–	–	–	–	–
Hydrogen	H_2	2.02	0.082	11.69	9.88	0.50	2.38	–	1.00	1.88	1.88	2.88
Methane	CH_4	16.04	0.656	36.45	32.81	2.00	9.52	1.00	2.00	7.52	8.52	10.52
Ethane	C_2H_6	30.07	1.229	63.85	58.40	3.50	16.67	2.00	3.00	13.17	15.17	18.17
Propane	C_3H_8	44.10	1.802	90.79	83.52	5.00	23.81	3.00	4.00	18.81	21.81	25.81
n-Butane	C_4H_{10}	58.12	2.376	117.72	108.64	6.50	30.95	4.00	5.00	24.45	28.45	33.45
Water gaseous	H_2O	18.02	0.736	–	–	–	–	–	–	–	–	–
Hydrogen sulphide	H_2S	34.08	1.393	22.91	21.91	1.50	7.14	1.00 (SO_2)	1.00	5.64	6.64	7.64
Atmospheric nitrogen	*	"28.17"	1.151	–	–	–	–	–	–	–	–	–

*i.e. Air without oxygen (nitrogen plus traces of argon, CO_2, etc.). The molecular weights are set down to 2 decimal place accuracy. In a previous table they were "rounded" off.

Weights of Products of Combustion

Combustible	Calorific value MJ/kg (@ 25°C)		Required for Combustions (kg)		kg Products of Stoichiometric combustion/kg						
	Gross	Nett	O_2	air	CO_2	SO_2	CO	N_2	H_2O	total (dry)	total (Wet)
C (to CO_2)	32.79	32.79	2.67	11.51	3.67	–	–	8.84	–	12.51	12.51
C (to CO)	10.12	10.12	1.33	5.74	–	–	2.33	4.41	–	6.74	6.74
CO	10.10	10.10	0.57	2.46	1.57	–	–	1.89	–	3.46	3.46
H_2	141.85	119.81	8.00	34.48	–	–	–	26.48	9.00	26.48	35.48
H_2S	16.45	15.14	1.41	6.09	–	1.88	–	4.68	0.53	6.56	7.09
S (to SO_2)	9.24	9.24	1.00	4.31	–	2.00	–	3.31	–	5.31	5.31
CH_4	55.54	50.00	4.00	17.24	2.75	–	–	13.24	2.25	15.99	18.24
C_2H_6	51.90	47.47	3.73	16.08	2.93	–	–	12.35	1.80	15.28	17.08
C_3H_8	50.34	46.31	3.64	15.68	3.00	–	–	12.04	1.64	15.04	16.68
C_4H_{10}	49.52	45.70	3.59	15.46	3.03	–	–	11.88	1.55	14.91	16.46
LPG propane	50.00	46.30	3.65	15.73	3.00	–	–	12.08	1.65	15.08	16.73
LPG butane	49.30	45.80	3.56	15.34	3.02	–	–	11.78	1.54	14.80	16.34
Gas oil	45.6	42.8	3.35	14.44	3.14	0.03	–	11.09	1.18	14.26	15.44
Heavy fuel oil	42.9	40.5	3.20	13.79	3.07	0.11	–	10.59	1.02	13.77	14.79
Industrial coal*	26.7	25.5	2.05	8.84	2.42	0.08	–	6.79	0.47	9.29	9.76

*Coal of 8% ash; only 0.92 kg per kg goes into combustion products.

Typical Properties of Commercial Fuels
(all volumes and calorific values are at 298.15K and 101.325kPa)

Fuel	Stoichiometric combustion requirements		Calorific value		Volume of products of stoichiometric combustion in air						"CO_2"* in stoich. dry products	Target values for good combustion		Density of Fuel
	O_2	air	Gross	Nett	CO_2	SO_2	H_2O	N_2	Total (dry)	Total (Wet)		O_2	CO_2	
Gaseous	m^3/m^3		MJ/m^3 at 25°C		m^3/m^3						%	%	%	kg/m^3
Natural gas (North Sea)	2.048	9.751	37.32	33.65	1.04	negl	2.02	7.72	8.76	10.78	11.9	2.1	10.7	0.72
Commercial propane	4.990	23.76	89.98	83.21	3.01	negl	3.96	18.77	21.78	25.74	13.8	2.1	12.4	1.86
Commercial butane	6.283	29.92	117.71	109.11	3.87	negl	4.82	23.64	27.51	32.33	14.1	2.1	12.7	2.38
Liquid	m^3/kg		MJ/kg		m^3/kg						%	%	%	g/cm^3
Kerosine	2.604	12.40	46.5	43.5	1.75	negl	1.71	9.79	11.54	13.25	15.2	3.7	12.5	0.77
Gas oil	2.561	12.19	45.6	42.8	1.75	0.01	1.60	9.63	11.39	12.99	15.4	3.8	12.7	0.82
Heavy fuel oil	2.453	11.68	42.9	40.5	1.71	0.05	1.39	9.22	10.98	12.37	16.0	4.0	13.0	0.95
Solid	m^3/kg		MJ/kg		m^3/kg						%	%	%	g/cm^3
Industrial coal rank 701	1.567	7.466	26.75	25.50	1.34	0.02	0.63	5.88	7.26	7.89	18.4	4.9	14.4	1.35
Industrial coke	1.698	8.088	27.90	27.45	1.66	0.01	0.17	6.40	8.07	8.24	20.6	4.9	15.9	1.10
Wood	0.905	4.312	14.35	15.80	0.87	negl	0.79	3.40	4.27	5.06	20.3	5.0	15.5	0.85

*The "CO_2 in dry products" is calculated as inclusive of any SO_2

Diagram Showing Mean Specific Heats of Various Gases

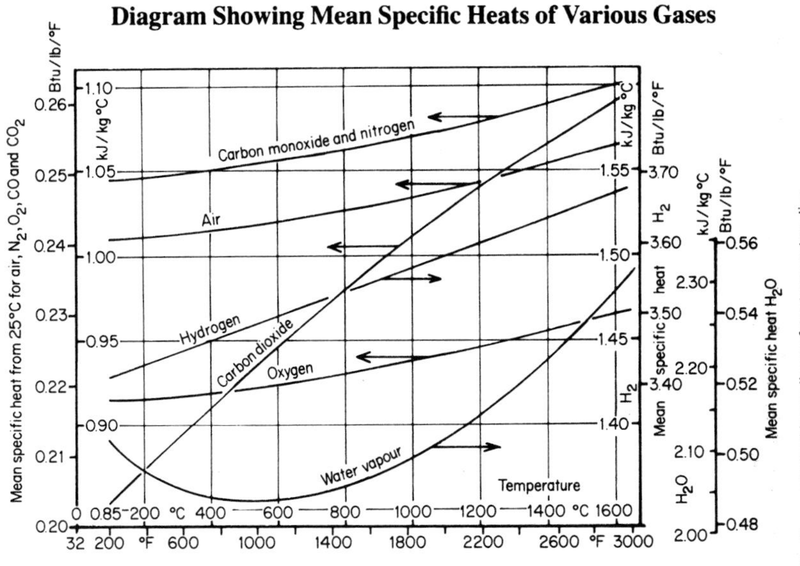

Flue Gas Losses When Firing Heavy Fuel Oil

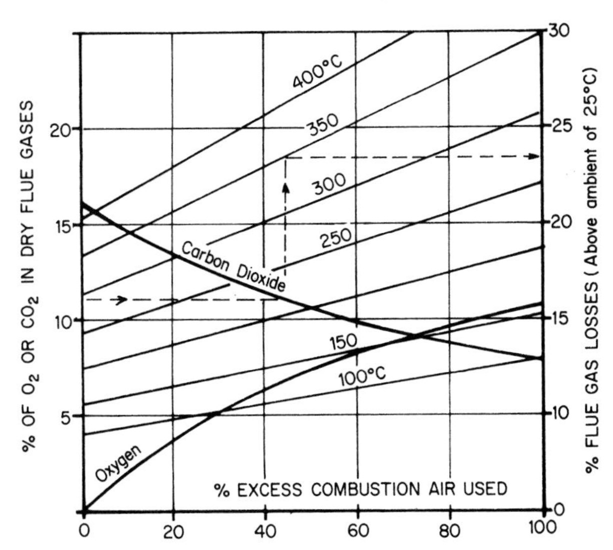

The lower diagram on page 35, and the next two allow a rapid estimation of the heat losses in products of combustion of the three fuels, heavy fuel oil, natural gas and coal. The diagrams are used by measuring the CO_2 or O_2 content of the exit gases and their temperature. Find the value of O_2 (or CO_2) percentage on the left-hand scale and move horizontally to the O_2 (or CO_2) curve. From the point of intersection move vertically up or down to the temperature curve, interpolating as necessary for the actual measured temperature. Then move from this second point of intersection horizontally to the right-hand scale and read off the percentage flue gas loss. The example–shows that for a furnace where measured CO_2 was 11% and exit gas temperature 350°C, the flue gas losses were 23% of the heat input. All three diagrams give heat loss above 15°C.

The diagram on page 38 combines the equivalent O_2 and CO_2 curves for four common fuels. There is an advantage in measuring O_2 in the exit gases from plant fired with two or more fuels since it will be seen that roughly the same O_2 reading is obtained with any of these fuels for the same excess air usage, but the CO_2 readings are very different. Thus for any plant where the fuel may be changed frequently, or even two fuels used simultaneously, control by O_2 measurement is preferable.

Flue Gas Losses When Firing Natural Gas

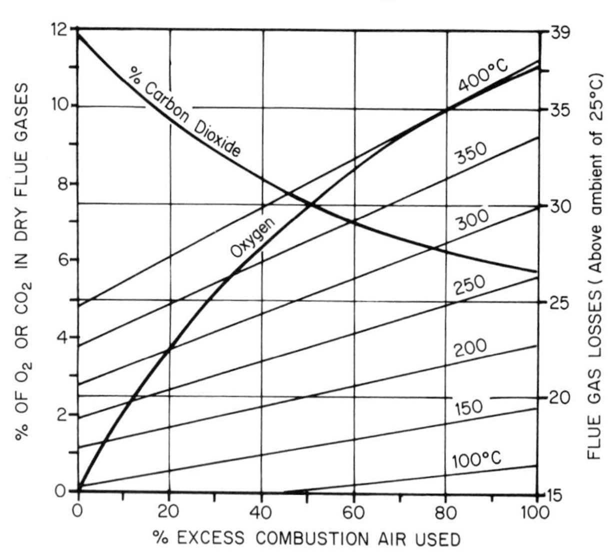

Flue Gas Losses When Firing Bituminous Coal

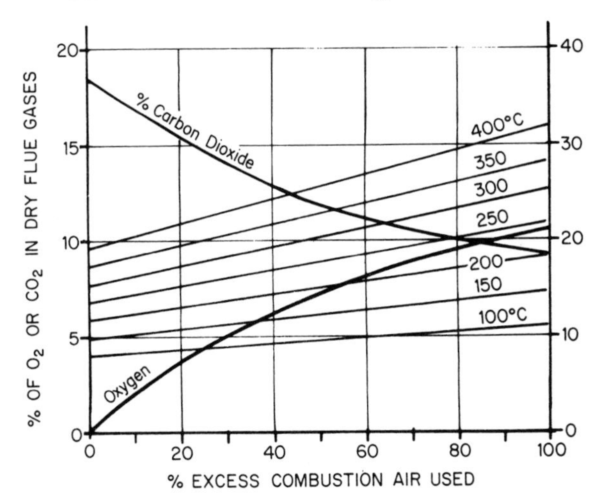

See previous page for significance of these two diagrams and method of use to estimate flue gas exit heat losses from a boiler, furnace or drier if the exit temperature and either the O_2 or CO_2 content of the exit gases can be determined. The diagrams also enable the excess combustion air usage to be estimated from the O_2 or CO_2 reading.

Combustion Chart for Typical Industrial Fuels

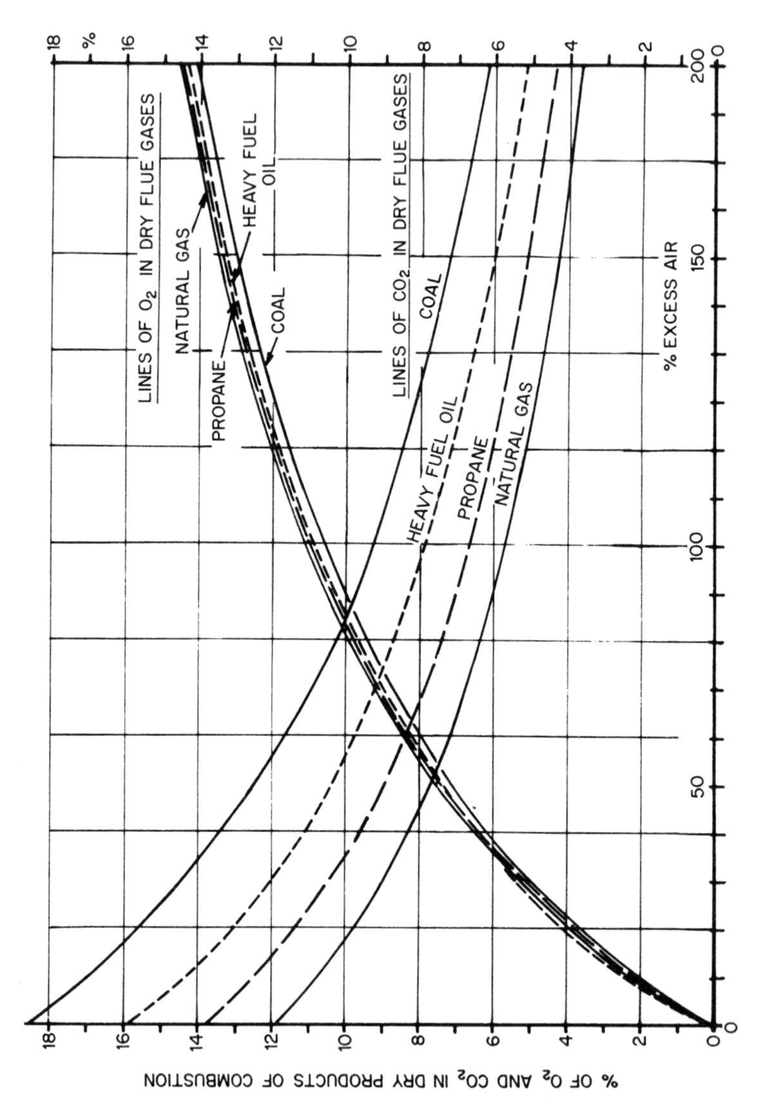

This combines the 3 previous diagrams and also adds the curves for propane, one of the liquefied petroleum gases, to demonstrate that the oxygen present in the dry products of combustion is related almost wholly to the excess air usage and only slightly influenced by the actual fuel. For example with an 8% O_2 reading, the excess air usage only varies from 54% for natural gas to 61% for coal.

5. HEAT TRANSFER, HEAT LOSSES

Measurement by Optical Pyrometer

The optical pyrometer reading may be inaccurate if the emissivity of the surface at which it is aimed is less than unity, and in many cases this will be so. A correction can be applied by the expression :

$$\left(\frac{1}{T_2} - \frac{1}{T_1}\right) = 1.045 \times 10^{-4} \text{ (arithmetical value of } \log_{10}E\text{)}$$

where T_1 is temperature read from pyrometer, T_2 is actual temperature, both in K (i.e. °C + 273.15) and E is the emissivity. The emissivity is normally less than unity, so its logarithm to base 10 will be −1.XXXX, and for the above equation it must be converted to its arithmetical value. (Example: Emissivity is 0.8. Log_{10} is bar $\bar{1}.9031$, so arithmetical value = −0.0069)

Some pyrometers have an adjustment scale to allow values of emissivity to be set and to make a correction roughly similar to this formula. Some values of E, the emissivity, are :

Surface	E
Newly Polished Metal surfaces	0.1
Aluminium Sheet oxidised surface	0.2
Stainless Steel Sheet after long use	0.6
Copper Sheet oxidised	0.6
Molten Metal Surfaces (exposed with any slag removed)	0.2–0.3
Refractory brickwork in boilers and furnaces	0.4–0.8
Brickwork and other rough matt external surfaces of buildings or furnaces at lower temperatures	0.7–0.9
Lime-washed or flat emulsion painted surfaces	0.8
Oxidised or rusty steel surfaces	0.4–0.7
Surfaces with gloss oil paint finishes	0.8–0.9
Surfaces with aluminium or bronze metallic paint finish	0.5–0.7
Glass	0.9

Heat Transfer by Radiation

The heat lost from a hot surface by radiation is calculated from :

$$Q = CE\,(T_s^4 - T_A^4)$$

where T_s is the temperature (K) of the hot surface, T_A is the surrounding temperature (K) to which the heat is being radiated, C is the Stefan-Boltzmann constant and E is the emissivity of the hot surface. Q is the heat loss per hour from each m^2 of hot surface. The value of C is dependent on the units in which Q is to be expressed e.g. :

$C = 5.67 \times 10^{-8}$ gives Q in watts/m^2
$C = 1.73 \times 10^{-9}$ gives Q in Btu/ft^2

See rough values for E in notes on Optical Pyrometer. *Caution:* slight differences in surface finishes can affect values of E so treat all values quoted with caution. There is also a variation in value of E with temperature of hot surface.

Emissivity of Flames

Heat transferred from flame to a surface almost wholly by radiation, i.e. varies as $(T_{flame} - T_{surface})^4$, but also affected by emissivity; both of surface, i.e. the absorption of heat depends on condition etc. of surface, and also emissivity of flame. Flame luminosity (which is main influence on emissivity of flame) largely influenced by ratio of carbon to hydrogen in fuel. Taking the various common industrial fuels:

Carbon/Hydrogen ratio:			
	Natural gas	3.0	(% by weight)
	35sec fuel oil	6.4	
	Heavy fuel oil	7.8	
	Coal	16.4	
	Creosote Pitch	14	(approx. depending on source)
	Creosote Oil	12	(approx. depending on source)

The following diagram shows typical flame emissivities for these fuels. Creosote pitch and coal are assumed burnt in suspension, e.g. by pulverising and firing through pf burner, so that comparison is fair. Conclusion is that fuels of low C/H ratio yield lower radiant heat transfer outputs — on a given shell boiler this means flue-gas temperature at end of furnace tube in which flame develops can be much higher than for a fuel with high C/H ratio. If boiler is loaded near to rating this could cause overheating of rear tube plate and smoke tube ends with consequent high repair bills, if boiler converted from say heavy fuel oil to natural gas.

Showing Effect of Carbon/ Hydrogen Ratio on Flame Emission

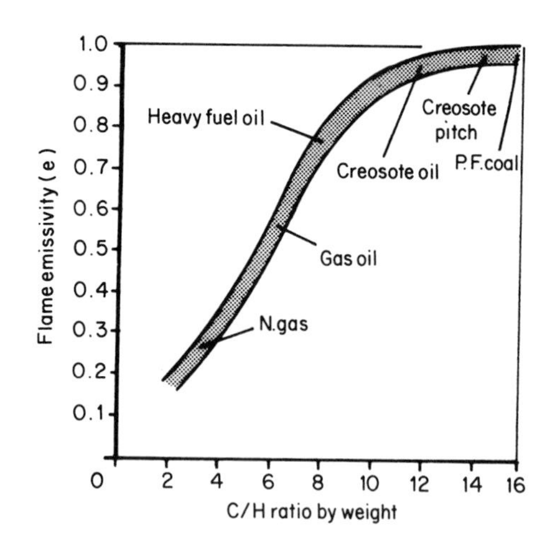

Heat Transfer by Natural Convection

The basic equation is rate of heat transfer in W/m^2, $W = C\,(T_1 - T_2)^{1.25}$, where T_1 and T_2 are the higher and lower temperatures in °C and the value of C varies with the shape and position of the surface. Typical values of C are :

2.56 for upward facing horizontal hot surfaces, or downward facing cold surfaces.

1.97 for flat vertical surfaces at least 0.5m high, or the curved surfaces of large diameter vertical cylinders, or an overall value for a large body of spherical or irregular shape with no re-entrants or cavities.

1.32 for downward facing hot surfaces.

2.30 for horizontal cylinders such as pipes, billets or ducts above 150mm diameter (this value increases rapidly as diameter decreases, examples being 3.9 for 10mm diameter, 13 for 1mm diameter, and 20 for 0.3mm diameter).

With forced convection, where the velocities over the surfaces are increased beyond those due to natural buoyancy movement of heated air or gas no very simple equations can be given. A very rough rule is that heat transfer varies as $V^{0.8}$ where V is the velocity across the surface, i.e. doubling velocity only increases heat transfer 1.75 times. However, some figures can be quoted for overall heat transfer in typical practical applications, allowing for radiation, convection, surface effects, etc.

Overall Coefficients of Heat Loss from Horizontal Bare Pipes

Take "still air" values for pipes inside buildings; 4km/h wind speed for pipes outside buildings in calm weather and the 16 and 32 km/h wind speeds for pipes outside buildings in "breezy" and "windy" conditions. The figures quoted are heat losses in W/m^2 °C, taking the outside area of the pipe and apply to dry weather. The losses could be doubled with steady moderate rainfall.

Nominal Pipe Size	*Temp. difference °C*	*0.5m*	*0.1m*	*25mm*	*12mm*
In Still Air	50	10.8	12.5	13.6	15.3
	100	13.6	14.8	16.5	18.2
	150	15.9	17.6	19.3	21.0
4km/h wind (2.5 mph)	50	16.5	18.2	19.9	22.1
	100	19.3	21.0	23.3	25.0
	150	22.7	24.4	27.3	29.0
16km/h wind (10 mph)	50	25.0	27.3	30.1	33.5
	100	29.5	31.8	34.6	37.5
	150	34.1	36.3	39.2	42.6
32km/h wind (20 mph)	50	30.7	34.1	38.6	42.0
	100	36.3	39.7	44.3	48.3
	150	42.0	45.4	51.1	54.5

OTHER OVERALL COEFFICIENTS are roughly as follows, in W/m^2 °C

Conditions	*Free convection*	*Forced convection*
Liquid to liquid heat exchangers – calorifiers water/water	140 – 340	850 – 1700
Liquid to liquid heat exchangers – oil heaters with hot water	28 – 55	110 – 280
Liquid to gas – hot water radiators Gas to liquid – economisers, waste heat recovery Gas to boiling liquid – boilers Condensing steam to air – steam pipes, air heaters	6 – 17	12 – 55
Condensing steam to water – condensers	280 – 1100	800 – 4500
Condensing steam to oil – oil heaters	60 – 170	110 – 340
Condensing steam to boiling water – evaporators, boiling pans	220 – 500	500 – 1700
Steam coils in water tanks	550 – 1700	—
Steam coils in oil storage tanks	85 – 140	—

Financial Value of Heat Losses from Uninsulated Saturated Steam Mains

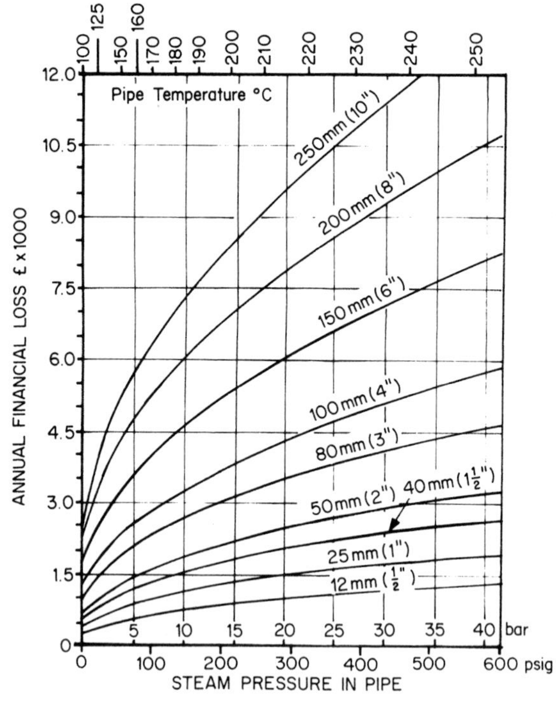

This diagram, and the one following, allow a quick estimate to be made of the financial losses occurring per year due to uninsulated steam mains or due to leaks. The financial values are correct for fuel costing 25 pence per therm gross calorific value (which is typical of a good quality coal at the end of 1984). These values also assume an annual average boiler efficiency of 78% on gross CV (for steam leaving the boilerhouse) and a continuously operating plant, with steam mains hot and under working pressure for 8700 hours per year.

The heat loss diagram can be used for pipes carrying pressurised hot water or organic fluids if the pipe surface temperature is known, by using the temperature scale on the top horizontal axis.

For other fuel costs or different annual running hours, correction factors are shown on the following page. These are used by multiplying the figure of annual financial loss obtained from either diagram, first by Factor A and then by Factor B.

Losses due to Steam Leaks via Holes etc.

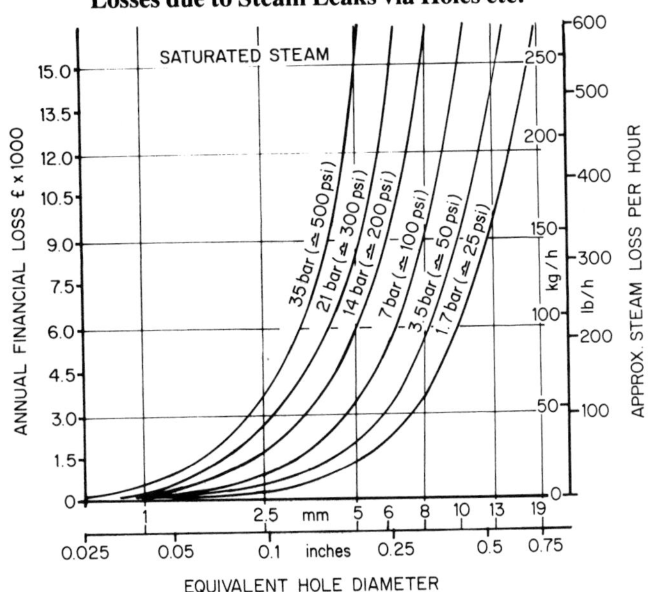

Factors to be used for other fuel costs or different total working hours

Factor A (Fuel)	Coal	= 1.0 (assuming 25p/therm)
	Natural Gas	= 1.28 (assuming 32p/therm)
	Light Fuel Oil	= 1.88 (assuming 47p/therm)
	Heavy Fuel Oil	= 1.56 (assuming 39p/therm)
	Propane	= 2.00 (assuming 50p/therm)

(any other price for fuel, multiply financial loss by (Cost/therm ÷ 25))

Factor B (hours)	Continuous	= 1.0 (8700 hours per year)
	Three-Shift	= 0.92 (8000 hours per year, comprising 333 working days, 24 hrs each)
	Two-Shift	= 0.48 (4200 hrs/year comp. 5 days/week with mains under pressure 17 hrs/day, 3 wks of annual shutdowns)
	One Shift	= 0.25 (2200 hrs per year, 5 days/week, 9 hrs/day, 3 wks annual shutdown)
	Day Space Heating	= 0.21 (1870 hrs/year, 5 days/week, 11 Hrs per day, 34 wks heating season.)

(any other working hours, multiply financial loss by (actual hours ÷8700))

Diagram for Estimation of Heat Loss Due to Unburned Carbon in Ashes

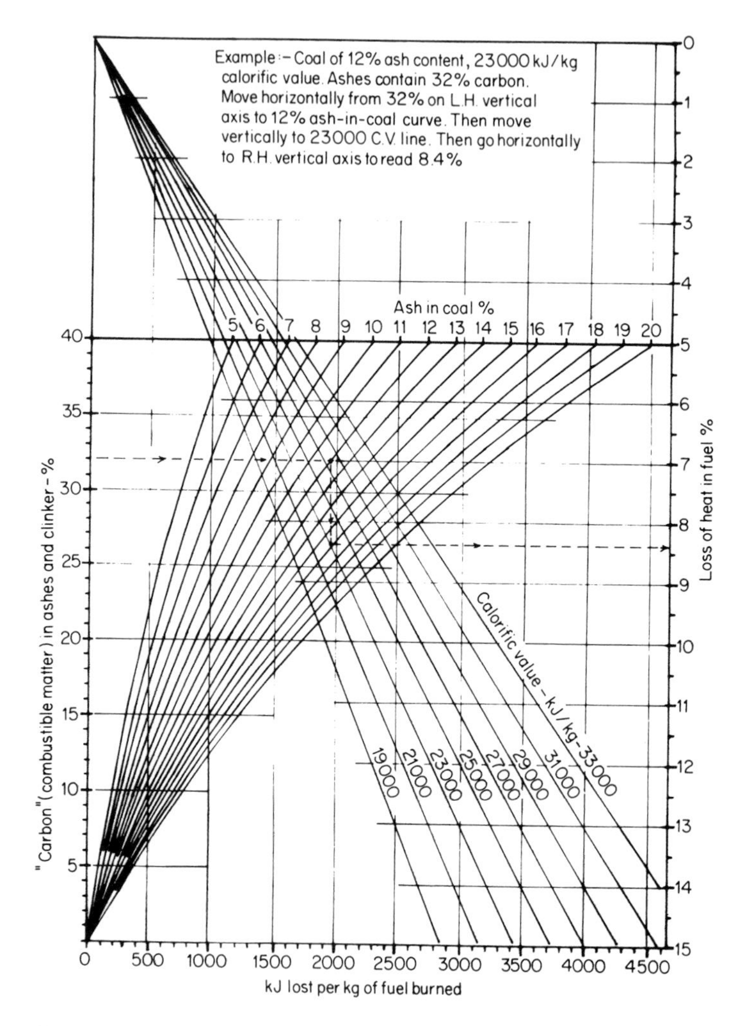

Example: Coal of 12% ash content, 23,000 kJ/kg calorific value. Ashes contain 32% carbon. Move horizontally from 32% on LH vertical axis to 12% ash-in-coal curve. Then move vertically to 23,000 CV line. Then go horizontally to RH vertical axis to read 8.4%.

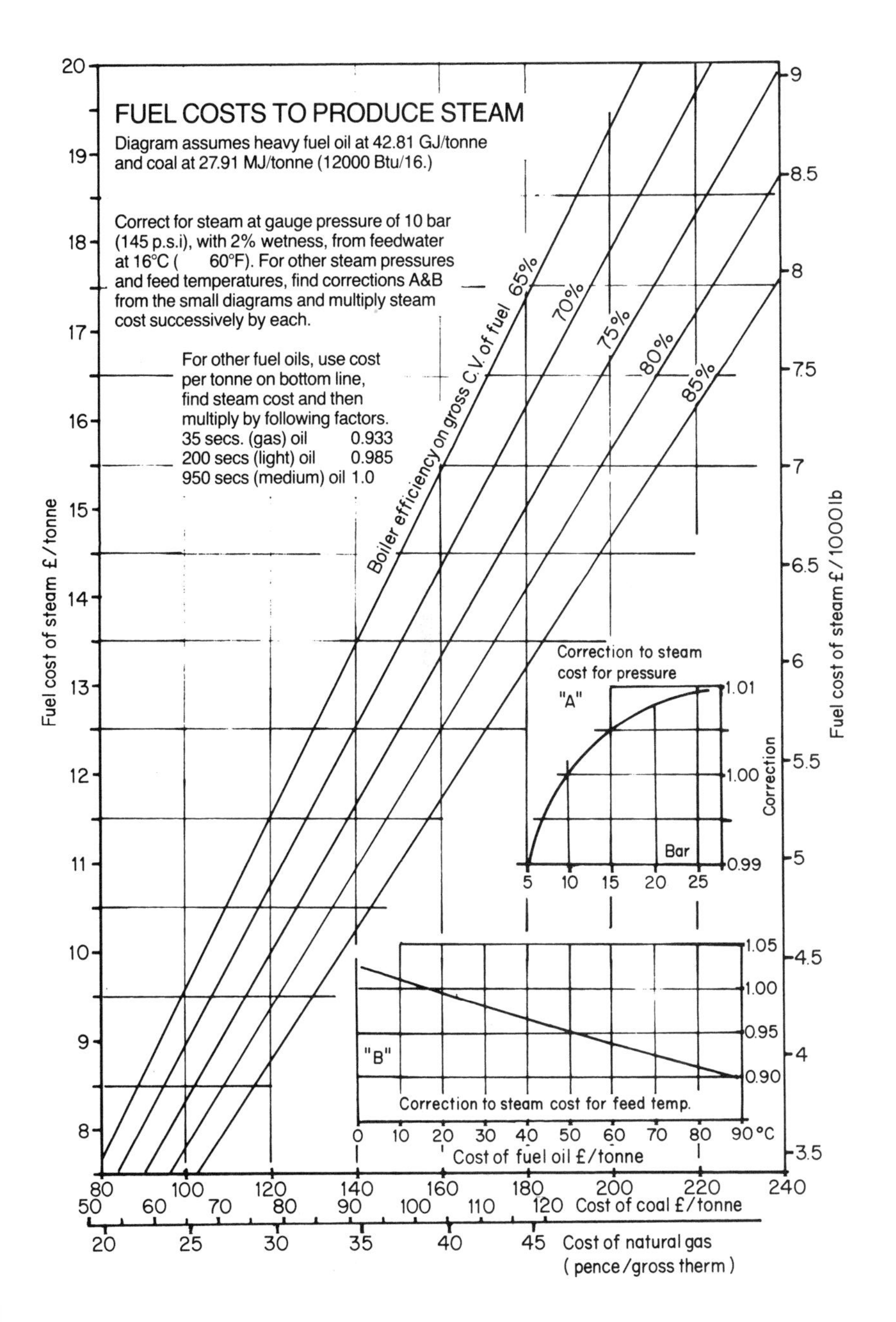
FUEL COSTS TO PRODUCE STEAM
Diagram assumes heavy fuel oil at 42.81 GJ/tonne and coal at 27.91 MJ/tonne (12000 Btu/16.)
Correct for steam at gauge pressure of 10 bar (145 p.s.i), with 2% wetness, from feedwater at 16°C (60°F). For other steam pressures and feed temperatures, find corrections A&B from the small diagrams and multiply steam cost successively by each.
For other fuel oils, use cost per tonne on bottom line, find steam cost and then multiply by following factors.
35 secs. (gas) oil 0.933
200 secs (light) oil 0.985
950 secs (medium) oil 1.0
Boiler efficiency on gross C.V. of fuel
65%
70%
75%
80%
85%
Fuel cost of steam £/tonne
Fuel cost of steam £/1000lb
Correction to steam cost for pressure
"A"
Correction
Bar
"B"
Correction to steam cost for feed temp.
Cost of fuel oil £/tonne
Cost of coal £/tonne
Cost of natural gas (pence/gross therm)

6. HEAT RECOVERY

A Review of Heat Recovery Devices

Device	*Description*	*Typical applications*	*Comments*
Gas to gas heat exchangers			
Plate heat exchangers	Gas streams separated by parallel or dimpled plates mainly in cross flow form. Materials frequently aluminium or coated aluminium but paper and stainless steel also.	Ventilation and Lower Temperature Process System.	Handle large volumes at low pressures. No cross contamination and good efficiency. Mostly applied below 200 °C but some units applied up to 600 °C. Retro fitting can be difficult.
Heat pipes	A sealed tube contains a wick, lining the inside wall, plus a working fluid. Opposite ends of tube are positioned in heated and cooled gas streams and heat is transferred by evaporation/cooling.	Process Systems.	Normally limited to max. temp of 400 °C but higher in some cases. Efficiencies only fair (about 60%). Fairly high costs.
Run around coil	Liquid circulated in closed circuit pipe system to coils positioned in supply and exhaust gas streams.	Ventilation and Lower Temperature Process Systems.	Low capital cost and often suited to retrofitting. Normally applied below 200 °C and operating efficiencies normally between 40 and 60%.
Shell tube heat exchangers	Gases flowing through tubes give up heat to gas flowing through surrounding shell.	Process Systems.	Often made on a one off basis to suit particular requirements. Bulky.
Gas to liquid heat exchangers			
Waste heat recovery boilers	Similar to conventional boilers but mostly without first radiant heat transfer pass. Designed for lower gas entry temperatures than conventional boilers hence larger sizes.	Process and Incineration waste gases.	Gas entry temperatures normally above 400 °C and sizes above 200 kW. Clean gases required to obviate frequent cleaning.
Flue economisers	Finned or plain tube bundles in boiler flue or process gas streams. Heat recovered as hot water and units may have integrated bypass.	Boiler and Process Waste Gases.	Wide range of sizes covering most commercial and industrial boiler applications. Can suffer from cold spots leading to corrosion
Fluidised bed	Gases pass through a shallow fluidised bed and heat hot water or steam in finned or plain tubes.	Fouled gas streams up to 1000 °C.	Compact and self cleaning but not yet fully developed.
Liquid to liquid heat exchangers.			
Shell tube heat exchangers	Liquid flowing through tube bundle conveys heat to surrounding liquid contained in a shell.	Process heat recovery.	Efficiency limited by tube heat transfer coefficients but available for wide range of temperatures and pressures. Low cost, simple

Liquid to liquid heat exchangers (cont.)			
Plate heat exchangers	Liquids separated by flat or dimpled plates to give contra flow heat exchange between the two fluids.	Process Heat Recovery.	Maximum temperatures below 500 °C and maximum pressures about 20 bar. Temperatures can be limited by the seals available to suit the fluid.
Recuperators (Also see diagram of possible savings by preheating air)			
Recuperative burners	Flue gases used to preheat burner combustion air. Recuperator may be integral with burner or separately positioned.	Furnaces and Processes above 600 °C.	Clean gases required to avoid fouling and/or corrosion of the heat exchanger. Materials limit applications to a maximum of about 1400 °C.
Spray recuperators	Clean non-corrosive exhaust gases in contra-flow with sprayed water to give hot water heat recovery.	Boiler and Process flue Gases.	Important that gases are clean, otherwise water is contaminated and secondary heat exchanger must be used. Good heat transfer. Simple.
Falling cloud recuperator	Exhaust gases in contra-flow with solid particles which fall to a fluidised bed heat exchanger to provide steam or hot water.	Hot dirty gases.	A new device now being proved. Should provide heat recovery to dirty gases.
Regenerators			
Heat wheels	A permeable matrix in the form of a disc slowly rotates. Half of the disc is in a hot gas stream and half in a cooler gas stream. The gases pass through the matrix of the disc.	Ventilation systems and clean lower temperature process gases.	Compact dimensions and low pressure drop combined with high efficiency. Cross contamination occurs and initial costs can be high.
Regenerative burners	When burner is not firing, hot exhaust gases flow through the burner and heat a ceramic heat sink. When burner fires, combustion air is preheated in the ceramic heat sink.	High temperature process work.	Only just commercially available but promises high efficiencies and suitability to high temperature clean gases.
Refractory regenerators	Refractory lined ducts alternate as exhaust and combustion air ducts.	Higher temperature processes, many continuous.	A bulky form of heat exchanger but capable of use with dirty and/or corrosive gases. Normally forms a part of the appliance.
Heat pumps			
Vapour compression type	Closed circuit system in which vapour is compressed before condensing to release heat. Then liquid absorbs heat in an evaporator before returning to the compressor.	Space heating/air conditioning and drying.	Uses low grade heat and therefore applied to low temperatures (up to 100°C) sizes range from few kW to several MW. Compressor shaft power provided by either electric motor or gas engine. Waste heat from latter can be usefully employed.
Absorption type	As above but compressor replaced by circulating pump and heat used to vaporise a refrigerant/absorbent solution.	As above.	Lower operating and maintenance costs than above. High efficiency potential if heating input re-used

Fuel Savings by Preheating Combustion Air

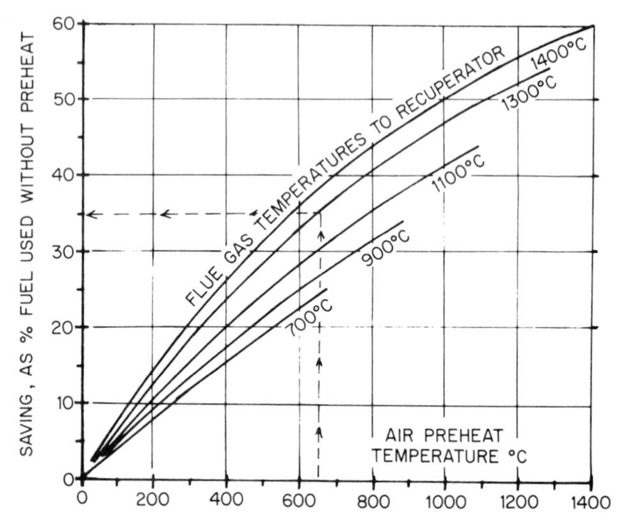

Diagram is for stoichiometric combustion requirements for natural gas.
Example: with stoichiometric firing and flue gases entering recuperator at 1300°C, preheating air to 650°C will save 35% of the fuel required without preheat

Recuperators may be of the following materials:

(a) Metallic convection type. Limiting temperatures are 600 to 800°C for gases and 500°C for air.

(b) Metallic radiation type. Limiting gas temperatures are 1350 to 1400°C but these can only be tolerated by limiting air preheat, to keep average metal temperature at a safe level. Expensive heat resisting steels must be used. Often combined with type (a), installed in series to take partly cooled gases.

(c) Refractory type. Uses refractory ducts or tubes. Rather bulky for amount of heat transferred and tend to suffer from leakage. Limiting gas temperature around 1200 to 1300°C with air preheats up to 700°C.

(d) Heat resisting glass – using glass tubes in gas stream – useful to resist certain corrosive gases, but max. gas temperature must be below initial softening temperature of glass.

The diagram shows "minimum" savings, as very few furnaces operate without excess air. Savings increase as essential excess air usage increases, for example:

Preheated Air Temperature °C	*Waste Gas Temperature 800°C – % fuel saved by preheat*		
	Stoichiometric	*10% Excess Air*	*20% Excess Air*
100	3.8	4.3	5.6
200	7.8	9.2	11.5
300	12.4	13.3	16.7
400	15.8	16.9	21.3
500	19.7	20.6	23.6

Heat Loss Due To Boiler Blowdown

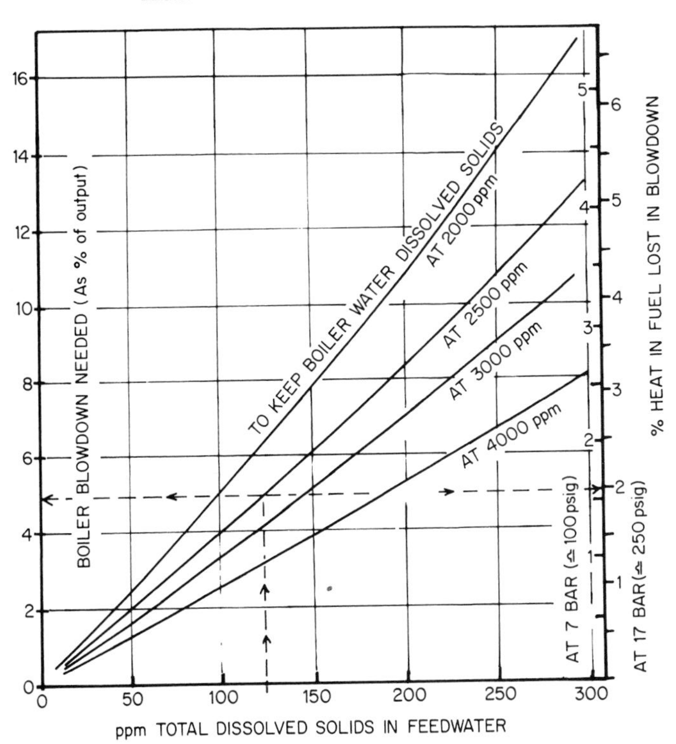

This diagram illustrates the amount of energy taken out of a boiler by the blowdown, whether intermittent or continuous, which is necessary to control the total dissolved solids in the boiler water. These must be kept below the level at which foaming or priming may occur and carry excessive moisture and dissolved solids over with the steam, and also to prevent rapid build-up of solids on the boiler heating surfaces. With current high cost of fuels it is now worthwhile to recover this heat. One method is to install a flash vessel to collect steam liberated when the pressure on the blowdown water is reduced, but more can be done. A heat exchanger can be installed to preheat cold make-up water going to the feed tank or cold water feeding the domestic hot water system for washing purposes, etc.

As an example, assume a boiler is fed with water containing 125 ppm dissolved solids. If the boiler water concentration is not to rise above 2500 ppm, a blowdown rate of 4.9% of steam output will be necessary. This blowdown contains 1.6% of heat in fuel used on the boiler, assuming a 7 bar operating pressure.

Chemicals Used in Boiler Water Treatment

Chemical		*Purpose*	*Comments*
Sodium hydroxide	NaOH	To raise pH to higher alkalinity to precipitate any magnesium.	
Sodium carbonate	Na_2CO_3	Again to raise pH also precipitate calcium	Safer to handle than caustic soda. Usually cheaper. Some carbonate breaks down to liberate CO_2 to steam output.
Sodium nitrate	$NaNO_3$	Used to inhibit caustic embrittlement	Was more essential for rivetted plates in boilers to control "cracking" round rivet holes etc.
Sodium aluminate	$NaAl_2O_4$	To precipitate calcium and magnesium and as a flocculating agent	Encourages depositions by forming a flocculent sludge.
Sodium sulphite	$Na_2 SO_3$	Used to eliminate oxygen in feed and boiler waters	Absorbs O_2 to form sulphate increases dissolved solids in boiler. Some decomposition at high temperatures to liberate hydrogen sulphide. Should not be used in boilers above 80 bar.
Hydrazine	$N_2 H_4$	To eliminate oxygen	Normally supplied commercially as solution around 35% strength. Advantage over sodium sulphite is no increase in dissolved solids and no sulphide formation. Very toxic. Handle very carefully. Must not be used near foodstuffs. On low pressure plant inject in feedline well before boiler.
Scale inhibitors	—	Control scale deposition by forming complex compounds with calcium and magnesium	Reduces deposition of scale on metal surfaces — need oxygen free water and should not be used at high pressures. One group called "chelants".
Polymers	—	Control scale deposition by keeping it as sludge	Prevents scale depositing on surfaces.
Tannins, seaweed derivatives, lignins, glucoses, etc.		Wide range of organic compounds used to keep scales in suspension or as loose sludges	Many sold as part of proprietary compounds. Can prevent hard scale formation and some can cause calcium and magnesium to form sludges.
Antifoams	—	Reduce tendency of boiler water to foam	Usually only useful where boiler water fairly high in dissolved solids — should be used only after trying higher blowdown rates.

Water Inside Boilers and as Feedwater

(A more detailed series of recommendations are available in B.S. 2486:1978)
Typical characteristics of water for various working pressures are as follows:

Boiler Water	*Shell Boilers*			*Water-tube Boilers*			
mg/litre	*Up to 10bar*	*15 bar*	*25 bar*	*Up to 20 bar*	*30 bar*	*45 bar*	*60 bar*
Total alkalinity as $CaCO_3$	400/ 1200	400/ 1000	400/800	max. 700	400/600	300/500	300
Caustic alkalinity as $CACO_3$	700	500	300	300	200	120	60
Chlorides (max.) as Cl_2	700	500	300	500	400	300	100
Silica as SiO_2	350	250	100	100	60	40	20
Sodium Phosphate as PO_4	40/80	35/70	30/60	30/60	25/60	20/40	12/30
Dissolved solids (max.)	3000	2500/ 3000	1500/ 2000	2000/ 3000/	1800/ 2200	1000/ 1500	800/ 1200
Feed Water to Boiler or Economiser mg/litre							
Total hardness (max)	10 mg/1 (less if possible)			10 at low pressure, decreasing to bare trace at 60 bar			

Feedwater to all types of boilers should have a pH value in range 8.5 to 9.5 and have no detectable traces of oil. Dissoved CO_2 and O_2 should be as low as possible, particularly for shell boilers above 15 bar, where O_2 value of below 0.05 mg/l should be maintained, and for water-tube boilers where O_2 should drop from 0.05 to 0.01 mg/l over the pressure ranges shown. In boilers themselves, a slight surplus of sodium sulphite or hydrazine should be maintained to ensure removal of any residual traces of O_2 — say 30 mg/l of Na_2SO_3 or 20 mg/l of hydrazine in 35% solution.

Boilers Out of Service

If a boiler is to remain unused for some time, certain precautions should be taken to prevent rusting and corrosion on the water spaces and tubes. The boiler should be completely filled right to the stop valve (including superheaters if fitted) but remember to have prominent notices — "Drain down to normal working levels before firing".

The water used to fill the boiler should be treated feedwater (see separate table for pH values etc.), but additional sodium phosphate should be added, allowing approximately 0.05kg/m^3 of total water capacity.

To prevent any significant oxygen attack, a scavenging chemical such as hydrazine or sodium sulphite should also be added, mixing this with a proportion of the water as it is added to top-up the boiler. Typical quantities are (in kg/m^3 of water):

	Either Hydrazine	*Or Sodium Sulphite*
For a shut down of 1 week	0.15 to 0.3	0.2 to 0.5
For a shut down of 1 month	0.3 to 0.7	0.5 to 1.0
For a shut down over 1 month	0.7 to 2.0	1.0 to 3.0

(For long shut down it is also satisfactory to drain the boiler completely and keep it perfectly dry — if possible close it up completely after inserting a tray of desiccant chemical to absorb water vapour — however, this is only satisfactory, if there is no air movement through the water spaces which could quickly saturate the desiccant and then possibly allow condensation to occur if the external surfaces of the boiler are exposed to Winter ambient conditions).

Solubility of Oxygen in Water at Various Temperatures

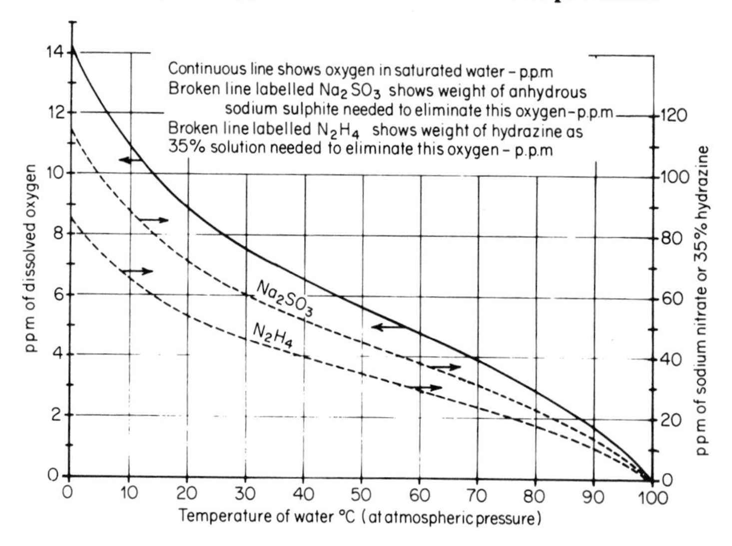

Financial Appraisal of Energy Saving etc. Projects

"Simple pay-back" — or the time it takes for the capital cost of a project to be recovered by savings (after allowing any extra labour, maintenance or other extra running costs) — is widely used to assess the financial benefits of any scheme such as an energy saving idea. Often this fails to allow for the long operating life of a project and future rises in the costs of fuels. Also it may not differentiate between two attractive projects when money is just not available for both, and it may not allow for inflation (£100 in 5 years time may not have the same purchasing power as £100 today).

"Discounted Cash Flow" introduces the concept that money has time value. A table of discount factors is required and a set (up to 15% discount rate and 15 years of future life) follows below, while page 54 gives an example, examining two different projects.

Discounting Factor Tables

Future years	Percentage rate of discount															Future years
	1	*2*	*3*	*4*	*5*	*6*	*7*	*8*	*9*	*10*	*11*	*12*	*13*	*14*	*15*	
1	0.990	0.980	0.971	0.962	0.952	0.943	0.935	0.926	0.917	0.909	0.901	0.893	0.885	0.877	0.870	1
2	0.980	0.961	0.943	0.925	0.907	0.890	0.873	0.857	0.842	0.826	0.812	0.797	0.783	0.769	0.756	2
3	0.971	0.942	0.915	0.889	0.864	0.840	0.816	0.794	0.772	0.751	0.731	0.712	0.693	0.675	0.658	3
4	0.961	0.924	0.888	0.855	0.823	0.792	0.763	0.735	0.708	0.683	0.659	0.636	0.613	0.592	0.572	4
5	0.951	0.906	0.863	0.822	0.784	0.747	0.713	0.681	0.650	0.621	0.593	0.567	0.543	0.519	0.497	5
6	0.942	0.888	0.837	0.790	0.746	0.705	0.666	0.630	0.596	0.564	0.535	0.507	0.480	0.456	0.432	6
7	0.933	0.871	0.813	0.760	0.711	0.665	0.623	0.583	0.547	0.513	0.482	0.452	0.425	0.400	0.376	7
8	0.923	0.853	0.789	0.731	0.677	0.627	0.582	0.540	0.502	0.467	0.434	0.404	0.376	0.351	0.327	8
9	0.914	0.837	0.766	0.703	0.645	0.592	0.544	0.500	0.460	0.424	0.391	0.361	0.333	0.308	0.284	9
10	0.905	0.820	0.744	0.676	0.614	0.558	0.508	0.463	0.422	0.386	0.352	0.322	0.295	0.270	0.247	10
11	0.896	0.804	0.722	0.650	0.585	0.527	0.475	0.429	0.388	0.350	0.317	0.287	0.261	0.237	0.215	11
12	0.887	0.788	0.701	0.625	0.557	0.497	0.444	0.397	0.356	0.319	0.286	0.257	0.231	0.208	0.187	12
13	0.879	0.773	0.681	0.601	0.530	0.469	0.415	0.368	0.326	0.290	0.258	0.229	0.204	0.182	0.163	13
14	0.870	0.758	0.661	0.577	0.505	0.442	0.388	0.340	0.299	0.263	0.232	0.205	0.181	0.160	0.141	14
15	0.861	0.743	0.642	0.555	0.481	0.417	0.362	0.315	0.275	0.239	0.209	0.183	0.160	0.140	0.123	15

Financial Appraisal of Two Projects

	Undiscounted				Discounted					
	Project A		*Project B*		*Project A*			*Project B*		
Year	*Per Annum* £	*Cumulative* £	*Per Annum* £	*Cumulative* £	*Per Annum* £	*Discount Factor At 10%*	*Present Value* £	*Per Annum* £	*Discount Factor At 10%*	*Present Value* £
0 • Capital Cost	(35,000)	(35,000)	(30,000)	(30,000)	(35,000)	1.0000	(35,000)	(35,000)	1.0000	(30,000)
1	10,000	(25,000)	15,000	(15,000)	10,000	.9091	9,091	15,000	.9091	13,637
2	10,000	(15,000)	15,000	0	10,000	.8264	8,264	15,000	.8264	12,396
3	10,000	(5,000)	15,000	15,000	10,000	.7513	7,513	15,000	.7513	12,270
4	10,000	5,000	15,000	30,000	10,000	.6830	6,830	15,000	.6830	10,245
5	10,000	15,000	15,000	45,000	10,000	.6209	6,209	15,000	.6209	9,145
6	10,000	25,000			10,000	.5645	5,645			
7	10,000	35,000			10,000	.5132	5,132			
Total Gross	70,000		75,000		Net Present Value (Total of RH column)		13,683	Net Present Value		17,693
Net	35,000		45,000							
Assumed Life		7 years	5 years							
Payback		3.5 years	2 years		Net Present Value		13,683			17,693
Gross Return on Capital		200%	250%							
Net Return on Capital		100%	150%							
Average Gross Annual Rate of Return		28.6%	50%		Internal Rate of Return		18% appx.			25% appx
Average Net Annual Rate of Return		14.3%	30%							

Note: Brackets indicate negative values

Example

Consider two projects. A has a capital cost of £35,000, will save £10,000 (reduction in fuel) per year after allowing various maintenance etc. costs and should have a life of 7 years. B costs £30,000, again save £10,000 per year, but will have a shorter life of 5 years. "Simple Payback" suggests A, at 3.5 years, less attractive than B at 2.0 years. "Discounted" techniques in the example below show B is indeed the better, but not by as wide a margin as first thought. (Even this can be further refined by altering annual savings by any predicted increases in real cost of fuels, by including effects of taxation on increased profits and by considering grants and capital allowances. These latter are being eliminated over the next few years). The net present value so obtained is the value in present day terms of all the income due to savings (less the initial investment) as they appear in the future over the estimated life of the plant. The NPV of B is shown to be greater than for A.

7. STEAM (AND HOT WATER)

Typical Costs of Raising Steam

Only too often new boiler plant is chosen on price – the cheapest reasonably reliable plant that can be installed. However, remember that even on single shift operation a boiler will burn more than its capital cost in fuel every year providing it is on a reasonable load. So it is the fuel bill that should be considered – not the capital cost! Any increase in capital cost that gives a year-round improvement in efficiency is usually one of the best investments any firm can make. The diagram shows typical costs of producing steam on heavy fuel oil and coal fired plants assuming reasonable efficiencies while actually working. On 1984 cost comparisons, a coal fired plant, even allowing for a capital cost possibly 1.5 to 2.0 times that of an oil fired plant, should be able to generate steam at about 70–75% of the cost on fuel oil.

One major factor affecting average annual steam costs is the boiler usage factor – if the plant only operates on a single shift basis there are considerable losses from the boiler etc. outside working hours and these have to be replaced by burning extra fuel. Also if the boiler is only lightly loaded it may be impossible to maintain good steady combustion – particularly with oil or gas burners which have a limit to turn down ratio. Stop/start operation with frequent purging, wastes a lot of heat.

The diagram on page 56 shows it is possible to have a rather oversized boiler, supplying a single shift winter heating load only, which would have a "usage factor" of only 10%, which would not produce steam below £17 per tonne, yet the same plant operating at the same combustion efficiency but on a year-round three shift process load above two-thirds boiler capacity could well have a reduced steam cost of £13 per tonne.

The diagram also shows that fuel costs can approach 80% of total operating costs (on oil, on single shift), once again emphasising that high all year-round efficiency is more important than capital cost. It can also be used for boilers producing hot water, taking 1 GJ of heat input to hot water (approximately 1 million Btu) as equivalent to the scale for cost of 1000 lb of steam.

7. STEAM (AND HOT WATER)

Cost of Steam or Hot Water From Typical Boiler Plants

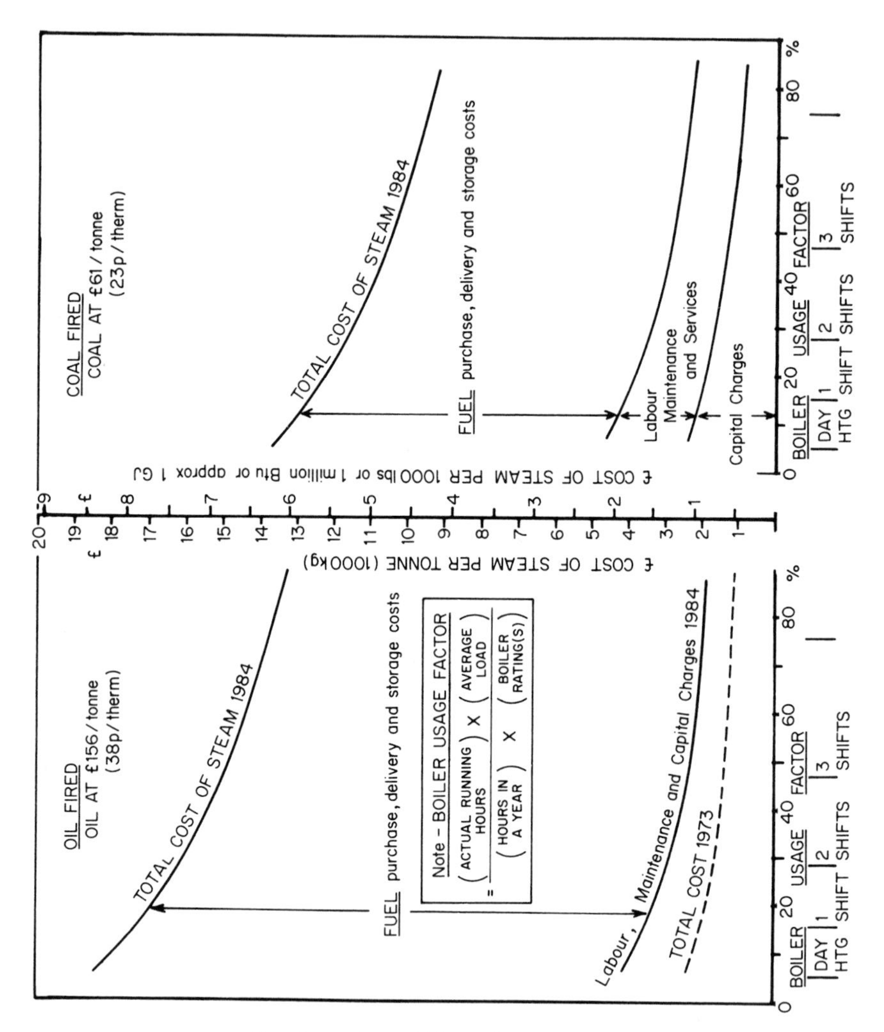

Condensate and Pumping Problems (Also applicable to any pumps handling hot water)

Condensate collection and return, and flash steam recovery to pre-heat cold make-up, all cost money and it makes economic nonsense if the hot water cannot be pumped. If the system cannot handle this hot water unless some is allowed to overflow to waste to allow cold water addition to reduce temperature, there is little point in insulating the condensate return lines!

Every system has an available net positive suction head (ANPSH) determined by temperature and the positions of tanks, pumps and pipe lengths. This ANPSH is the head hopefully available to drive water into the pump suction. Similarly every pump has a Required NPSH (RNPSH) which is the head needed to drive water into the pump to permit the desired delivery quantity to occur. If ANPSH is less than RNPSH, cavitation, flashing and loss of output will occur.

ANPSH = potential head – (loss in head due to vapour pressure of hot water and also deducting friction loss of water passing through pipes and fittings plus entry loss (velocity head) to accelerate water up to the required speed for the quantity needed entering the pump)

Potential Head = Atmospheric Pressure (as a water column at the pumping temperature, e.g. 1012 mbar = 10.4m of water at 10°C, 10.7m at 80°C or 10.8m at 100°C) and either *minus* suction lift in metres if feed tank below the pump or *plus* static head in metres if fuel tank above the pump.

The diagram on page 58 can be used to check any particular layout. Follow these examples on it.

(a) Pumps 0.6m above water in tank, water at 80°C. Calculated friction and velocity losses in pipes equivalent to 1.8m.

Enter top line at 80°C, drop to curve to give a correction for vapour pressure at 80°C and go horizontally to LH axis marked "loss in head due to vapour pressure". This gives a Head reading of 5.2m. Add the friction and velocity losses of 1.8m to give 7.0m. The potential head is atmospheric pressure water column (10.7m at 80°C) minus suction lift of 0.6m giving 10.1m. Therefore, ANPSH = (10.1–7.0) = 3.1m. If maker's curves for pumps show a RNPSH of 3.5m for the delivery pressure needed for this application then the system will not be satisfactory. The tank must be raised or friction losses reduced until ANPSH is 3.5m or better.

(b) Pump 2.5m below water in tank, water at 70°C, calculated friction and velocity losses 2.5m. Drop from 70°C to curve and horizontally to read 3.7m; add on the friction etc. loss again of 2.5m to get 6.2m. This time the potential head is 10.6m (atmospheric water column at 70°C) *plus* 2.5m water head above pump suction, or 13.1m. There can therefore be no problem since ANPSH = (10.6 + 2.5 – 6.2) = 6.9m.

Continuing this example; if maker states 3.5m is RNPSH there is a margin of (6.9 – 3.5) = 3.2m, i.e. the loss of head due to vapour pressure could increase by 3.2m from the 3.7m (70°C) to (3.7m + 3.2m) = 6.9m. If this figure is run across from the LH axis, it meets the curve at a point corresponding to 88°C, so that for this example all should be well for water temperatures up to 88°C.

All levels refer to the water level in the tank – always take the lowest likely level – do not assume that condensate return, make-up water, etc. will always enter at such a rate as to maintain the tank at maximum water level. Always design the suction piping between tank and pump to give minimum friction resistance – use swept fittings, easy bends, gate valves, etc. and generally at least one pipe size larger than actual suction correction of pump to minimise velocities.

Vapour Head Corrections for Pumping Problems

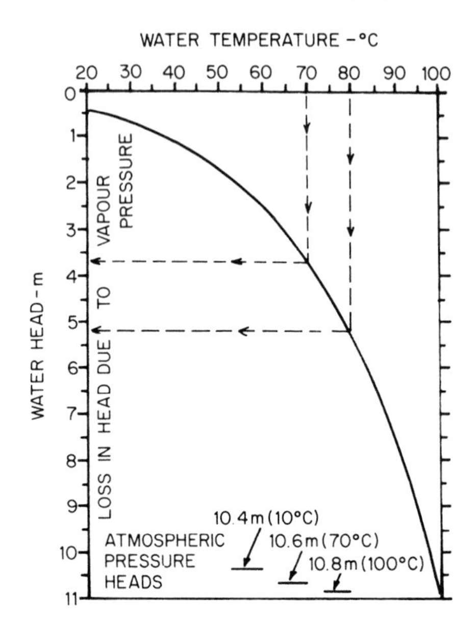

Combined Heat and Power

The diagram considers standard maximum electrical demands up to 2500kW and shows how various combinations of electrical generating set with waste heat recovery can be used to meet this load and also a process steam/space heating demand. To give complete comparability the heat output is standardised as steam at 3.5 bar (approximately 50 psig) but it must be realised that a greater amount of waste heat might be possible in certain cases if lower steam pressures, or hot water, can be used by customers since this could enable further cooling of exhaust gases.

Lines on the diagram are :

A. Diesel engine, with exhaust gas boiler. Feed water for boiler preheated to maximum possible temperature by use of heat exchanger on jacket cooling water.

B. Diesel as in A, but with larger exhaust gas boiler and an extended duct between engine and boiler containing an oil or gas burner which uses the hot exhaust gases, still rich in oxygen, in which to burn additional fuel and so increase steam output. Line shows maximum steam availability, using up nearly all oxygen in exhaust gases, but any steam output between lines A and B is obtainable by varying fuel to after-burner. Because of increased steam output/kW, most of the cooling jacket water can be used to preheat feedwater.

C. Gas turbine using natural gas or distillate fuel oil, with exhaust gases passing through waste heat boiler.

D. Gas turbine as in C, but with after-burner to fire additional fuel (which can be cheaper, i.e. heavy fuel oil) to increase output from boiler in similar fashion to B.

E. Steam turbine, taking in high-pressure steam at 17 bar (approximately 250 psig) and exhausting at 3.5 bar. This power/steam line is more rigid than for the diesel or gas turbine cases and there is no opportunity to boost low pressure steam production by after-burning.

Modern fuel prices and relaxations in the conditions applicable to private generation of electricity, can make some schemes financially attractive – they must however operate long hours, preferably three-shift all year round and it is preferable to carry the economic maximum electrical rating steadily with fluctuations met by the public supply. On a more modest scale consider the almost forgotten ideas of driving boiler fans and pumps by steam turbines – this can be financially attractive as long as all the exhaust steam can always be used for de-aeration, feed water heating, or process.

In passing, the relative efficiencies of the cases considered are :

% of Heat Input	*To Electric Power*	*In Exhaust*	*Cooling*	*Other Losses*	*Available for Heat Recovery*
Diesel set	38	36	21	7	57
Gas Turbine	18	74	–	8	74
Steam Turbine	9	83	–	8	83

When extra fuel is fired in exhaust gases of diesel or gas turbine generators to raise more steam in larger waste-heat boilers, the efficiency of usage of this extra fuel is very high – around 95%! This is because, compared to the use of a waste-heat boiler with no re-firing, the mass-flow of waste gases is only slightly increased by the weight of the extra fuel, and exit temperatures are similar, so nearly all the extra heat input goes into the boiler water and produces steam.

Relationship Between Possible Steam and Electrical Outputs for Various Generating Sets

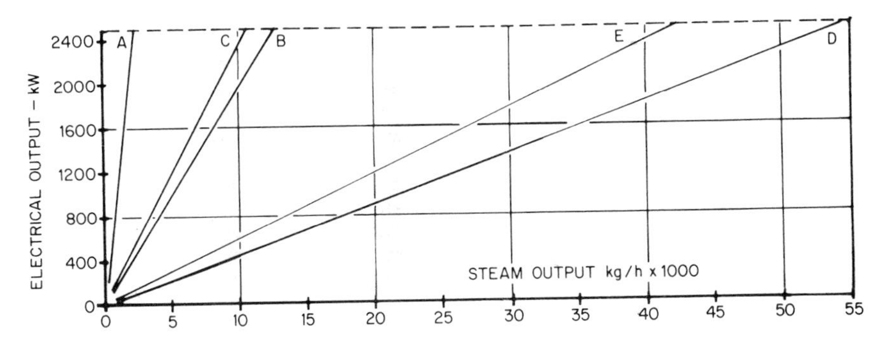

Power Generation by Steam Turbines

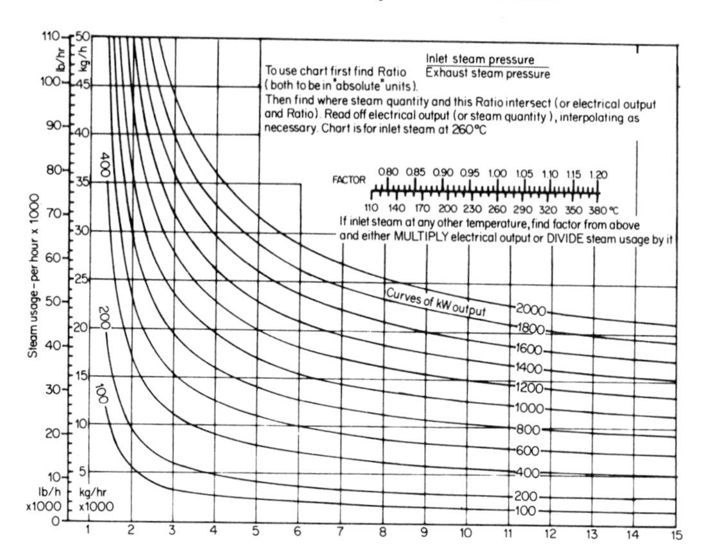

Linear Expansion of Pipes

Expansion is stated as mm increase in length per 100m of pipe, above 20°C.

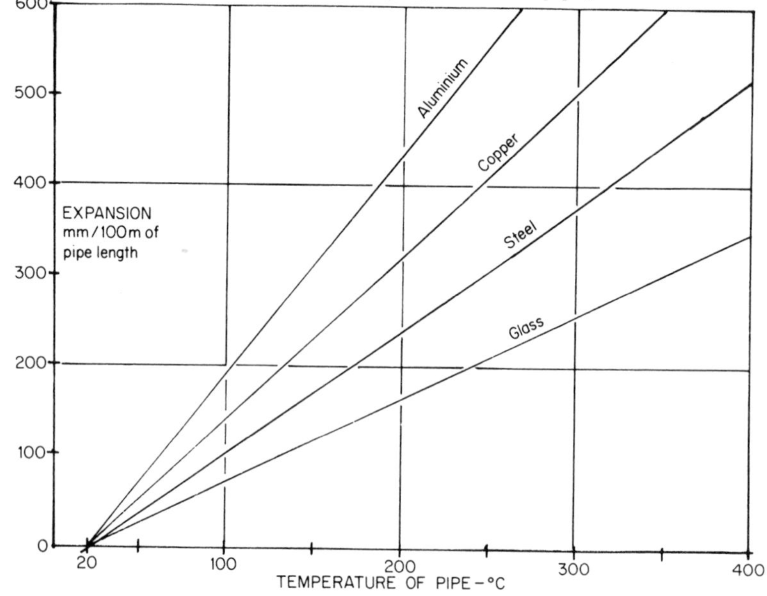

Power Output from Water Turbines

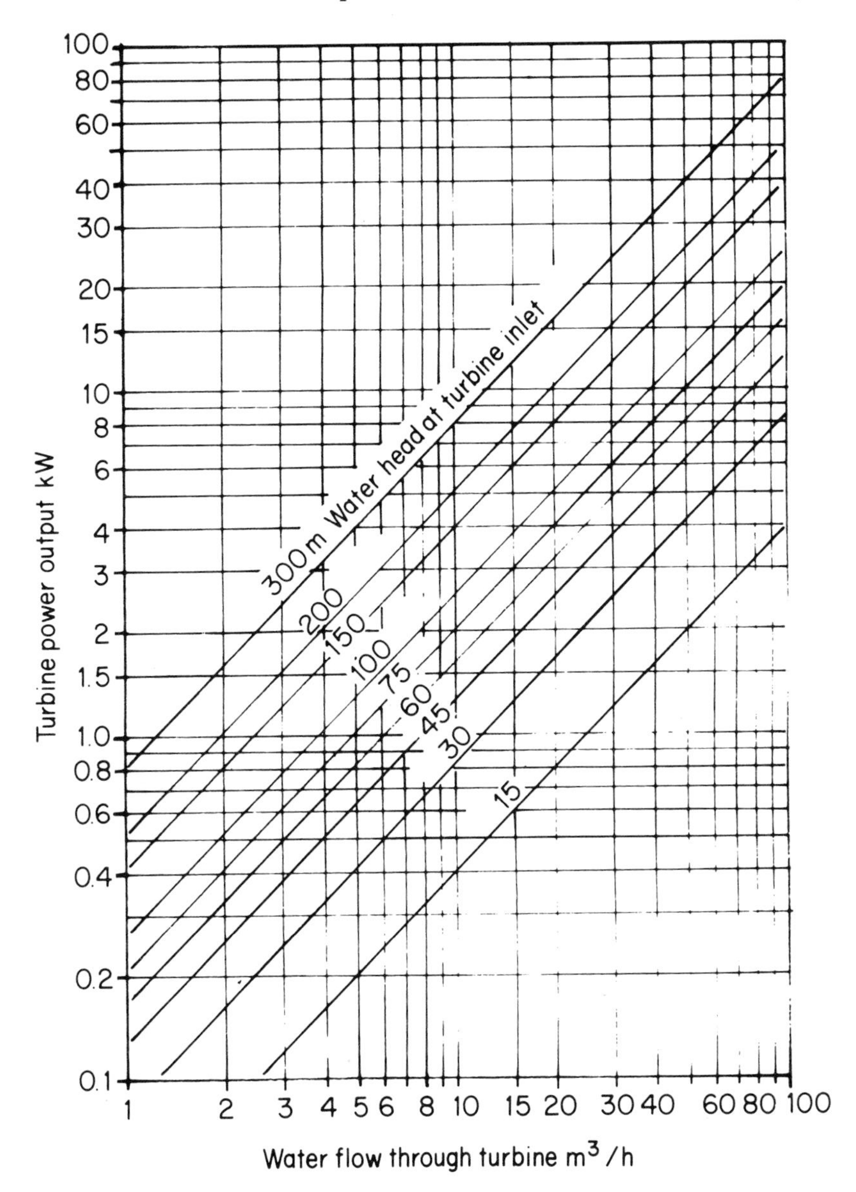

7. STEAM (AND HOT WATER)

Pressure Drop for Natural Gas in Fairly Clean Steel Pipework at Low Pressure (10 to 20 mbar; i.e. 4 to 8 in. H_2O)

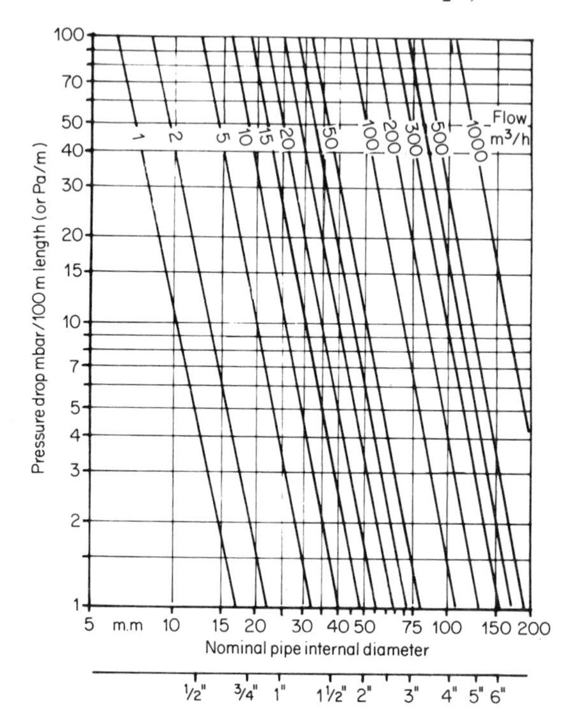

Allowance for Fittings, to be Added to Actual Pipe Lengths, as Equivalent Pipe Diameters

Valve – angle type	30	Small radius bend	15
Valve – gate	12	Large radius bend	10
Valve – globe	60	Elbow	25
		Tee – straight through	10
		Tee – to or from side branch	40

For example a 100mm globe valve (approx. 4″ pipe) would create a resistance to flow equivalent to 6 metres of straight pipe.

(These are average values, sufficient for many calculations, but actual allowances do vary with velocity. Where very accurate values of resistance to flow are needed, the CIBS Guide gives a detailed procedure.)

Recommended Economic Velocities in Pipes (m/sec)

Fluid	*Velocity*	
	Low Pressure	*High Pressure*
Water	1.5–2	3
Natural Gas } Air }	6–7	12–13
Wet Steam	20	25
Dry Saturated Steam	28–30	40–43
Superheated Steam	40	55

Miscellaneous

Water temperatures normally used:

Hot baths and shower baths	40°C (105°F)
Scalding hot	66°C (150°F)
for wash basin hot taps	46°C (115°F)
Swimming baths	27°C (80°F)
Hot water storage (hard water)	62°C (145°F)
Hot water storage (soft water)	80°C (175°F)
Cold mains water (winter)	4–5°C (40°F)
Cold mains water (summer)	16°C (60°F)

Approx. quantities of hot water normally used :
Normal bath – 12 gallons hot, 12 gallons cold
Shower bath – 5 gallons total (hot depends on mixed temperature)
Rough Estimate: 1kWh used via immersion heater heats 14 litres (3 galls) to 65°C (150°F), or 11 litres (2½ galls) to 80°C (175°F).

Flash Steam

When water above atmospheric pressure, and at a temperature corresponding to boiling point at that pressure, is released to a lower pressure, its temperature must drop to the lower boiling point. This causes part of the water to be converted to "flash steam" using the surplus heat energy to provide the latent heat of evaporation. The percentage of water converted to flash steam is usually shown as a diagram, but this rarely shows the heat value of the flash compared to the initial heat content of the original pressurised water. This can be considerable and so a Table has been constructed on page 65 to show both the flash steam by weight, and the heat content, compared to an original one kilogramme of water. As an extreme example, it will be seen that when boiling water at 15 bar is released to atmospheric pressure, 18.8% is liberated as flash steam and this steam contains 50.4% of the original heat content. Try to think in future of flash steam as a loss of a large proportion of heat rather than a minor proportion of weight.

Use the Table with caution – it really only applies to the types of steam traps that discharge condensate immediately it is formed. Some thermostat type traps retain condensate until its temperature is below boiling point at that pressure and with these less flash will be formed when the pressure is reduced. The Table can still be used if this temperature can be measured just before the trap. For example, if steam conditions are 10 bar (184°C) but the condensate is not discharged until it has cooled to 175°C the Table shows that with a receiver at atmospheric pressure only 13.9% by weight of flash steam will be formed, not the 15.6% that might have been assumed.

A similar reduction in "flash" can occur where due to bad design or overloading, "waterlogging" occurs due to the trap not discharging condensate fast enough at peak periods and the condensate is retained for some time and loses temperature by heat losses through pipework etc.

Remember the Table is for traps (and blowdown systems) maintained in excellent condition. If a steam trap has a worn valve seat, it may not shut tightly when condensate has been discharged and may allow the passage of steam, giving an artificially high apparent amount of "flash".

Either install a flash steam separating vessel, if the low pressure steam can be usefully employed, or cool the condensate via heat exchanger, air heater, etc. before it passes to any open vessel to remove more than enough heat to prevent flash formation – in effect, cool the condensate to 100°C or less if it is returning to an open-vented tank.

Flash Steam Formation (from condensate discharged under pressure or boiler blowdown)

Discharge conditions											
Pressure – bar	1	2	3	4	5	6	7	8	9	10	15
– psig	14.5	29	43.5	58	72.5	87	101.5	116	130.5	145	218
Corresponding temperature °C	120	134	144	152	159	165	170	175	180	184	202
Heat content of pressurised water above 0°C kJ/kg	505	561	603	636	666	691	712	733	754	770	844
Reduced to atmospheric pressure											
Heat released to form flash steam kJ/kg	86	142	184	217	247	272	293	314	335	351	425
% flash steam formed/kg	3.8	6.3	8.2	9.6	10.9	12.1	13.0	13.9	14.8	15.6	18.8
% of total initial heat in flash	17.0	25.3	30.5	34.1	37.1	39.4	41.2	42.8	44.4	45.6	50.4
Reduced to 1 bar											
Heat released to form flash steam kJ/kg	–	56	98	131	161	186	207	228	249	265	339
% Flash steam formed/kg	–	2.5	4.4	5.9	7.3	8.4	9.4	10.3	11.3	12.0	15.4
% of total initial heat in flash	–	10.0	16.3	20.6	24.2	26.9	29.1	31.1	33.0	34.4	40.2
Reduced to 2 bar											
Heat released to form flash steam kJ/kg	–	–	42	75	105	130	151	172	193	209	283
% flash steam formed/kg	–	–	1.9	3.5	4.9	6.0	7.0	7.9	8.9	9.7	13.1
% of total initial heat in flash	–	–	7.0	11.8	15.8	18.8	21.2	23.5	25.6	27.1	33.5
Reduced to 3 bar											
Heat released to form flash steam kJ/kg	–	–	–	33	63	88	109	130	151	167	241
% flash steam formed/kg	–	–	–	1.5	2.9	4.1	5.1	6.1	7.1	7.8	11.3
% of total initial heat in flash	–	–	–	5.2	9.5	12.7	15.3	17.7	20.0	21.7	28.6

7. STEAM (AND HOT WATER)

Steam Storage in Water Under Pressure

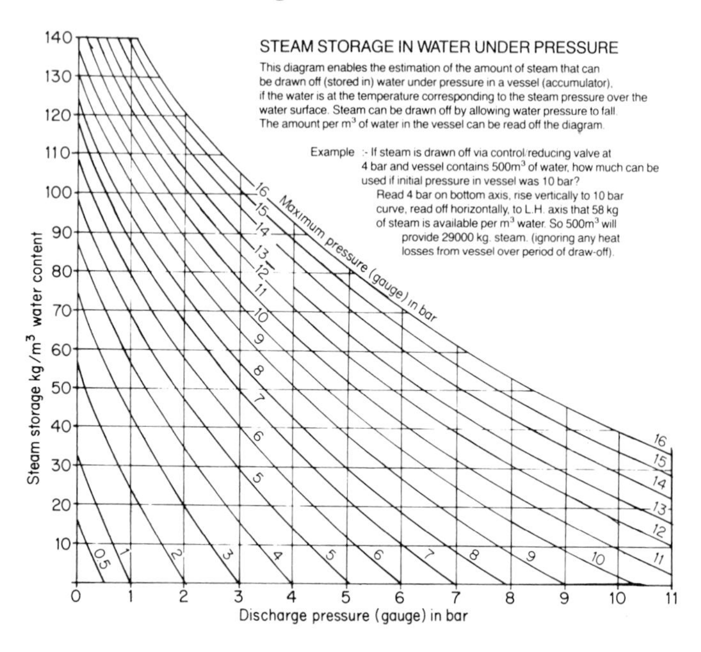

Flow

While the metering of steam is well known, using orifice plates, venturis or other fittings inserted in straight lengths of pipework, there are methods by which the flow of water, or other liquids can be checked using existing open tanks or troughs and where such measurements can then enable flow balances or energy balances to be constructed so that possible energy conservation schemes can be evaluated.

The following diagram is useful for two cases:-

(a) Where a horizontal pipe is partly filled with a flowing liquid. The wetted perimeter of the pipe, as a percentage of internal diameter can be obtained if the depth of liquid is measured. Similarly, the wetted perimeter of a circular hole cut in the side of a tank to act as an overflow can also be obtained.

(b) Where a horizontal cylindrical tank is partly full of liquid, the cross-sectional area of liquid, as a percentage of the area of the total circular cross-section, can be obtained. This enables the volume of liquid in the partly-full tanks to be obtained, and, if the density is known, the weight.

The nomogram on the following page enables the flow through such a circular orifice or pipe to be obtained once the wetted perimeter has been obtained. Where water is flowing through an open duct or out of a tank, a "V" notch weir can be used to measure flow, providing the flow out of the "V" has a clear fall. A table to allow a flow chart to be constructed for various included angles is given on page 69.

Diagram to Determine (a) Wetted Perimeter of Pipe or Orifice (partly filled)
(b) % of Area Occupied by Liquid

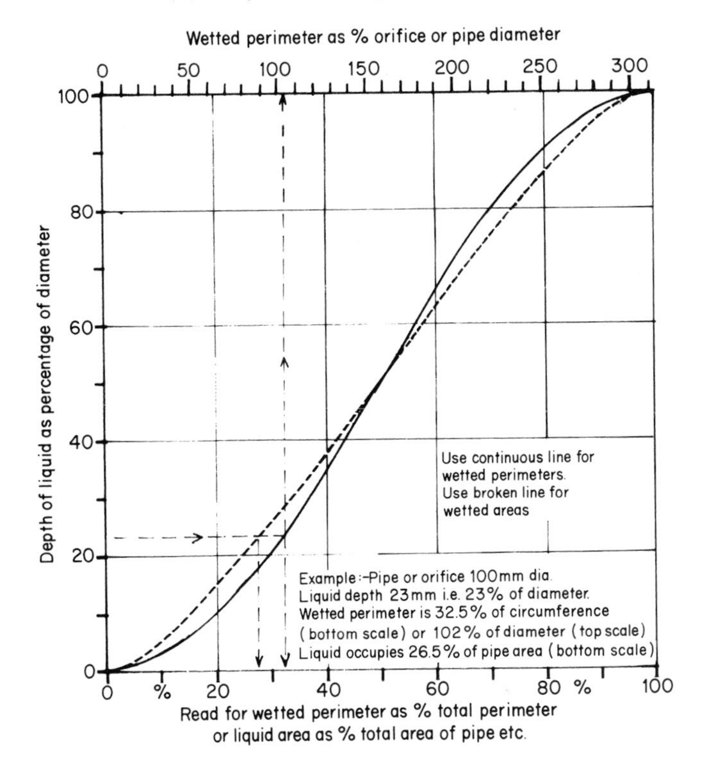

Diagram For Flow Over a Circular Weir

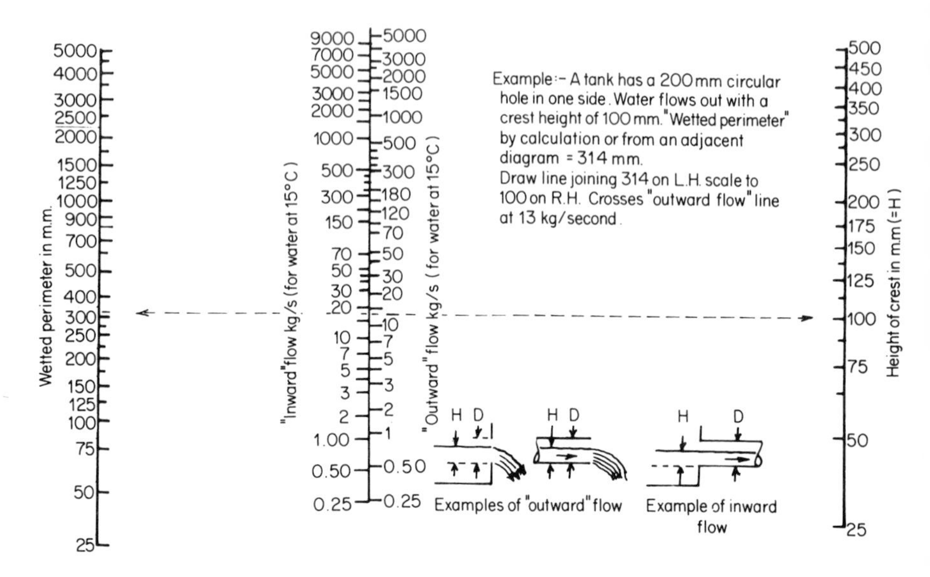

Permanent Pressure Drop Across an Orifice Plate

When steam or compressed air pass through a metering orifice plate there is an immediate pressure drop, of sufficient size to give an adequate differential pressure signal to a meter – meters might have "heads" of up to 0.6 metres of Hg (about 0.77 bar) and some engineers may feel this pressure change is a permanent loss of pressure and could be unacceptable. However, much of the signal pressure is recovered downstream of the orifice plate. Typical unrecovered pressure drops for an orifice sized to give a 0.42 bar pressure signal (6 psig) at full flow on a meter are as follows :

Ratio orifice diameter to pipe internal dia.	*% permanent loss of pressure*	*Loss in bar (psig) after partial recovery*
0.25	92	0.39 (5.6)
0.35	86	0.36 (5.1)
0.45	79	0.33 (4.7)
0.55	69	0.29 (4.1)
0.65	58	0.24 (3.4)
0.75	47	0.20 (2.9)

Water Flow Through "V" Notch Meter

The "V" must have a sharp up-stream edge, must be made either of thin metal or have the downstream side bevelled at approximately 45°, and the flow from the "V" must have a clear fall of at least 30mm before hitting any other solid or liquid surface. On the up-stream side the bottom of the "V" should be at least twice the anticipated maximum flow depth above the bottom of the tank. The height of the water flow through the "V" should be measured in the up-stream tank, not at the "V" itself.

Water height in mm	*Included angle of "V" – flow in m^3/h*					
	10°	*20°*	*30°*	*45°*	*60°*	*90°*
25	0.044	0.090	0.145	0.218	0.304	0.522
50	0.241	0.495	0.768	1.181	1.570	2.863
75	0.704	1.386	2.091	3.227	4.545	7.727
100	1.455	2.810	4.363	6.810	9.310	15.909
125	2.500	5.000	7.500	11.364	15.910	27.725
150	3.864	7.727	11.810	18.180	25.000	43.180

Flow Rate from a Horizontal Pipe

The diagram overleaf enables the flow rate of water from a horizontal pipe to be estimated. Measure the internal diameter "d" of the pipe and make sure the end is cut square and without any burrs or taper and there is no scale or other deposits in the pipe. Choose "a", a horizontal distance from the pipe end equal to one of the figures on the sloping lines on the left-hand diagram (100 to 2500 mm) and measure the vertical distance by which the water stream centre-line has dropped below the pipe centre-line, "b". Find the intersection of "b" with the chosen value of "a" on the left-hand diagram, move horizontally to the value of "d", drop to the bottom line and read off flow in litres/minute.
Example:- 50 mm pipe, flow drops 250 mm after 700 mm horizontal distance. Broken line shows how flow rate of 320 l/m is obtained.

Flow Rate from a Horizontal Pipe

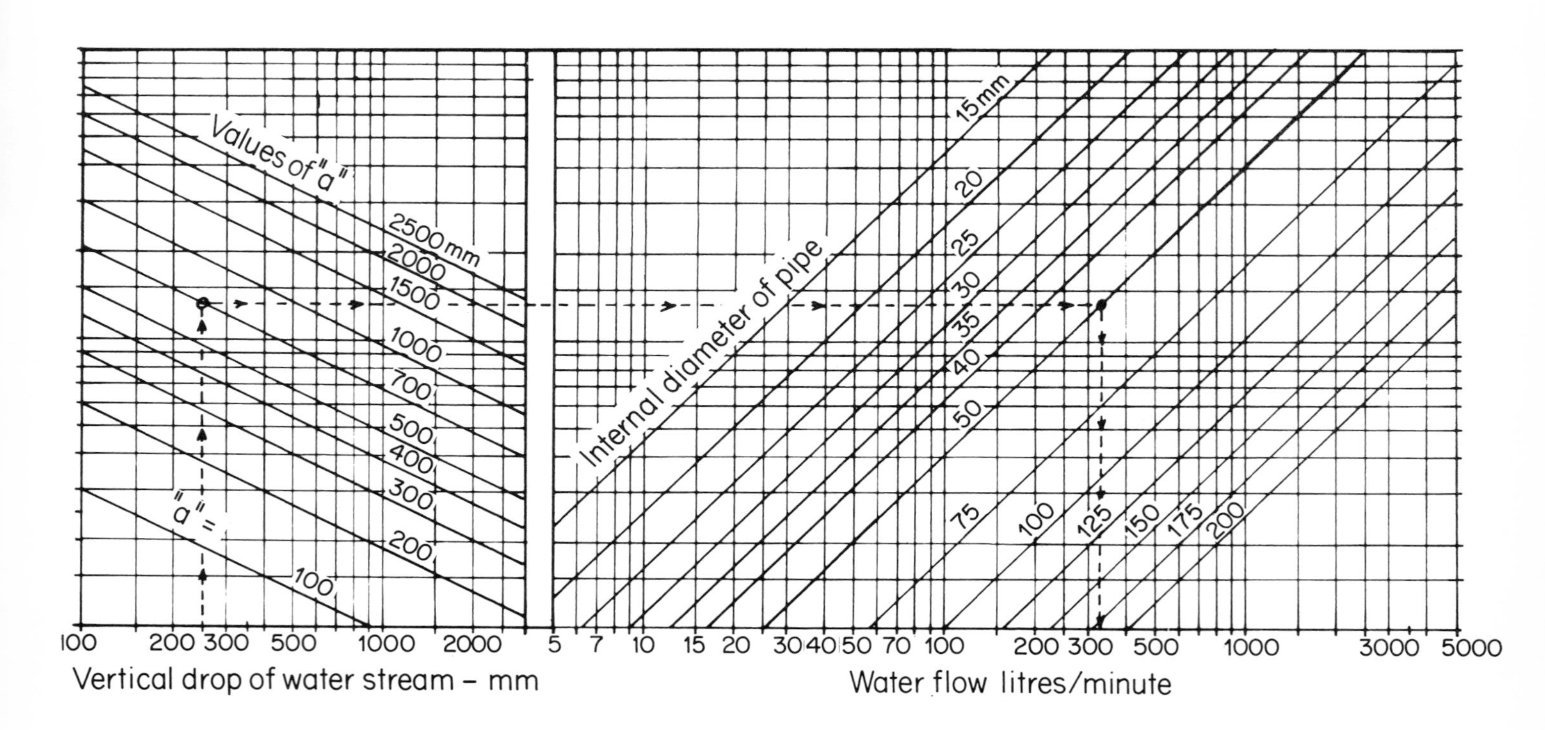

Superheat in Steam due to "Wiredrawing"

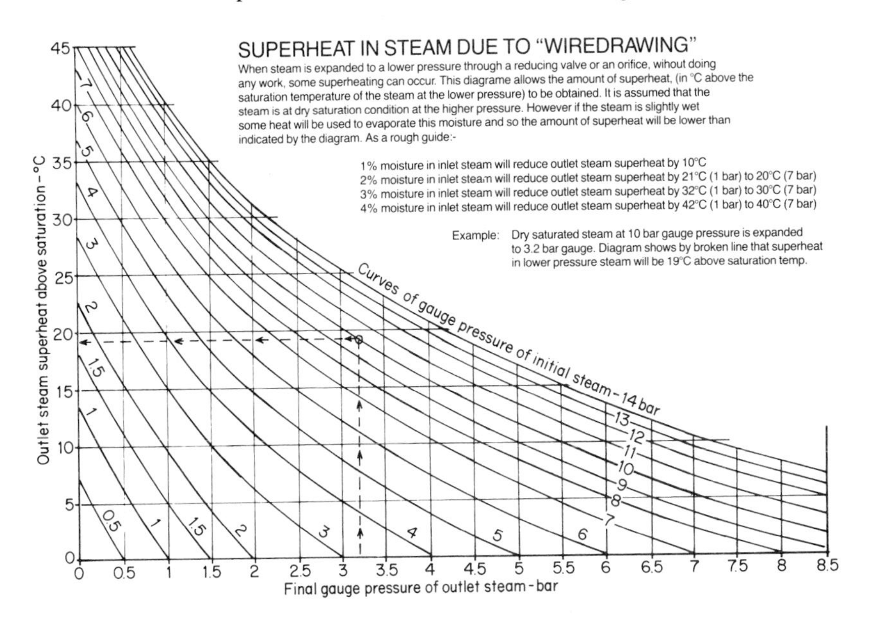

Correction Factors For Steam Meters

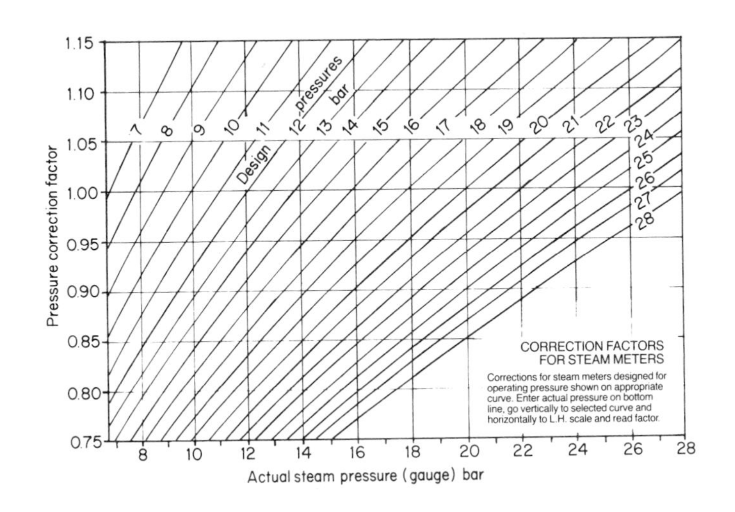

Correction Factors For Steam Meters

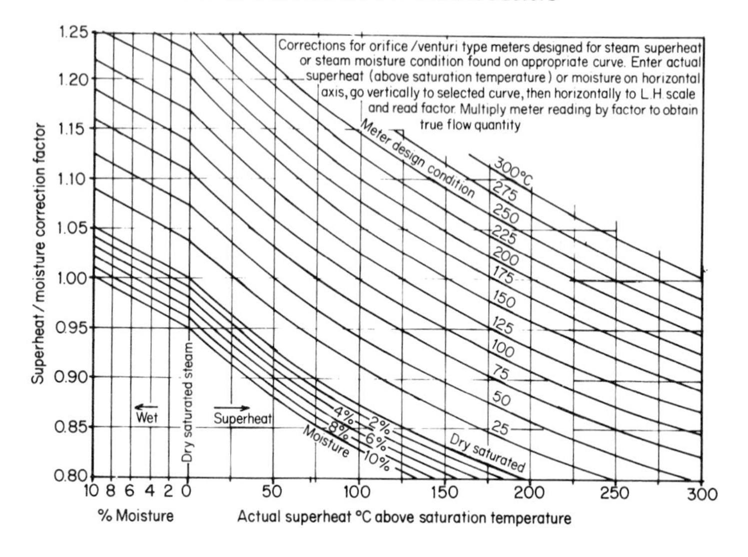

8. Heat Pumps and Refrigeration

Heat Pumps

First diagram shows principle. Working fluid under pressure is expanded through a valve which causes temperature to fall. Cold fluid passes to a heat exchanger ("evaporator") where it picks up heat, using this to supply the latent heat of evaporation, and becomes a vapour. This is compressed by the pump, and its temperature rises due to work done in compressing it, and then passes to a second heat exchanger ("condenser") where it cools, gives up its latent heat and condenses to pass as a liquid to the expansion valve to begin the cycle again.

Efficiency in transferring heat from one temperature to another called Coefficient of Performance.

Ideal CoP = $\dfrac{T_2}{T_2 - T_1}$ (where T_2 is evaporator temperature and T_1 is condenser temperature both in degrees absolute)

For example if evaporator temperature is 20°C (293K) and condenser temperature is 70°C (343K) the theoretical CoP is 343/(343—293) i.e. 6.86.

This ideal is never achieved; in practice a CoP of ⅓ to ½ ideal is quite good.

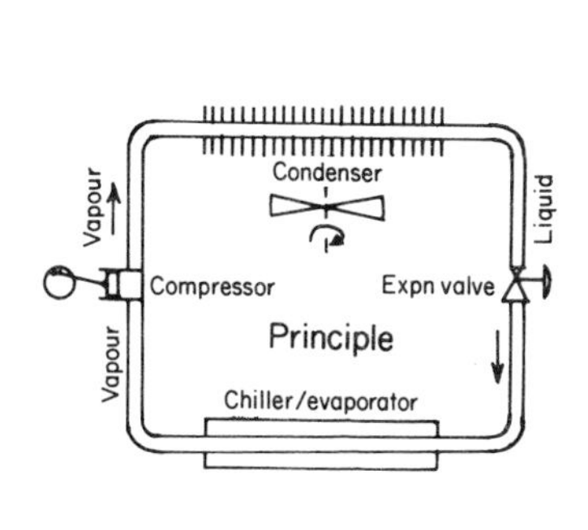

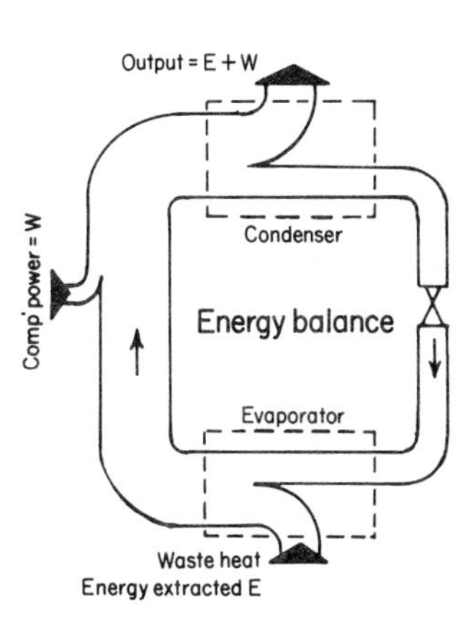

From an energy point of view we are interested in ratio $(E + W)/W$

Where E is heat transferred from evaporator to condenser (or more accurately from waste heat source supplying heat to evaporator to useful output) and W is power usage of pump. For example, an electrically driven pump taking 1kW of electrical power and having a practical CoP of 4 will deliver 4kW of heat from the condenser. The cost of this electrical power compared to providing the same heat via conventional fuels and a boiler or air heater will show whether the heat pump can be financially viable.

8. HEAT PUMPS AND REFRIGERATION

Example

1kWh of input electricity — cost say 3.8 pence on current tariffs.

Natural gas at 32 pence/therm; i.e. 1.09 pence/kWh

Gas used at say 65% overall efficiency in boiler/connecting pipework/calorifier to provide hot water; cost of hot water then 1.68 pence/kWh

If heat pump delivers 4kW of hot water per input kW of electricity operating cost = 0.95p/kWh of hot water

Saving is therefore 0.73p/kWh and must be evaluated against annual amount of hot water delivered and capital/maintenance costs of heat pump in excess of boiler system to determine rate of pay-back.

Sometimes the CoP is quoted as ratio of power delivered by compressor to compress vapour compared to useful heat produced — remember friction and cycle losses in compressor and efficiency of electric motor may demand 1.5kW input electricity to produce 1kW of work on vapour — so be careful to check how CoP is defined by manufacturers.

Practical Examples

1. Compressor can be driven by diesel or gas engines, steam turbines, or even gas turbines in large installations and this enables waste or exhaust heat to be recovered/recycled. If maximum use can be made of this, can give cheaper running costs than purchased electricity and electric motor drive.

Example

Gas engine drive (converted car or lorry engine)

Gas to engine — 1 therm, costing say 32 pence

Used as follows:

Shaft power	—29% used at CoP heat pump of 4.5	=	1.31 therms
Engine Cooling	—21% which should all be usable	=	0.21 therms
Engine exhaust	—40% of which 60% recoverable	=	0.24 therms
Other losses	—10% friction, other heat losses not recoverable	=	—
	Total useful ouput		1.76 therms
	i.e. 18.2 pence/output therm		

Shaft power (0.29 therm) is equivalent to 8.5 kWh and allowing for electric motor efficiency would require 10 kWh input to motor, costing 37.2p at current tariffs — and this would only provide 1.31 therms heat pump output. To uprate heat pump to 1.76 therms to equal gas engine scheme would require 13.4kWh input, costing 50p. So gas engine scheme about 64% running cost of equivalent electrically driven scheme. Another case of advantages of combined heat and power.

2. Diagram shows a possible application for steam boiler plants where proportion of feedwater is cold make-up. If there is a low temperature heat source nearby, such as cooling water from process or refrigeration, the heat pump can upgrade this to a higher temperature sufficient to justify preheating boiler feed. (Always make sure first that waste heat has been minimised before trying to make a case for the heat pump!)

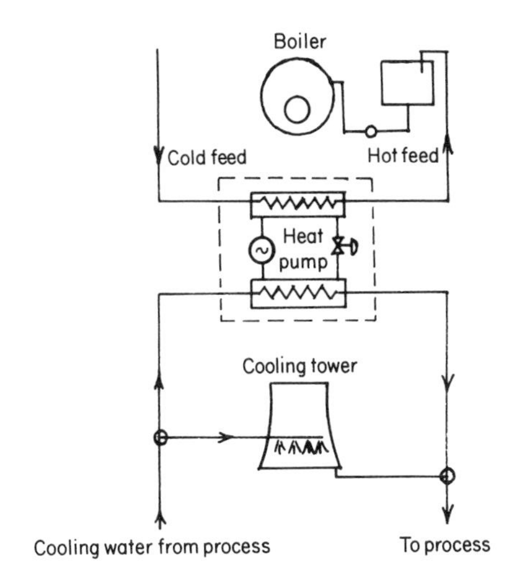

Efficient Running

Whether running as a heat pump or as a straight refrigeration system, the main items to check for efficient operation are:

1. The suction pressure between the evaporator and the inlet to the compressor
2. The discharge pressure between the compressor outlet and the condenser
3. The logbook of oil consumption (if the compressor is a lubricated type)

The two pressures should remain at maker's limited settings while the compressor is running, with only slight variations for changes in water or air temperatures entering evaporator or condenser. Oil added to lubricate compressor means oil has been lost into the system (the compressor is *not* like a motor car engine, it does *not* burn oil!). The table overleaf indicates possible faults:

Suction Pressure	
Low	*High*
Poor Refrigerant flow Shortage of refrigerant Process temperature below design Process duty being restricted	Too little superheat
Discharge Pressure	
Faulty compressor valves Shortage of refrigerant	Fouled condenser Fault in cooling tower Overcharge of refrigerant Air in system
Excessive Running Time	
If evaporator fan-cooled, is air distribution even over coils or are baffles damaged, causing air to move over only part of coil? Is there excessive icing on evaporator tubes? De-icing should be frequent and efficient. If heat rejected via cooling water and cooling tower, check tower performance. Poor sprays, damaged baffles, blocked air inlets, can all cause performance to fall off. Is system frequently purged of oil carry-over? Oil in circuit can foul heat exchange surfaces and increase pressure drops. If several systems in parallel, check minimum number in use for required load. Avoid excessive idle running.	

Refrigerants

The ASHRAE Standard Designation of refrigerants is based on an alphanumeric code that describes the molecular structure. For hydrocarbon compounds this code is the letter "R" followed by several numbers.

1st number = number of double valency bonds between atoms in the molecule
2nd number= number of carbon atoms in excess of the first
3rd number = number of hydrogen atoms, plus 1
4th number = number of flourine atoms in the molecule

If the first or second numbers are 0, i.e. no double bonds or only one carbon atom, these zeros are not printed.

Thus for dichlorodifluoro-methane we can state:
1st number = 0; 2nd number = $1 - 1 = 0$; 3rd number = $0 + 1 = 1$; and 4th number = 2. So code is R12.

Similarly for propylene ($CH_3CH = CH_2$), code is R1270.

If an isomer exists of the same chemical composition, a suffix "a" is used. For example, butane is R600 and isobutane is R600a.

For inorganic compounds used as refrigerants the ASHRAE code is 'R' (for refrigerant) followed by the figure "7" followed by the molecular weight of the gas. Ammonia (molecular weight 17) is R717. The system is not so logical where the refrigerant consists of a mixture of two gases. While it could be described by using the two numbers, one for each component, a smaller and more arbitrary number is given, a three figure number beginning with "5" — for example, a mixture of R12 (see above) and R152a could be described as R12/152a, but instead is arbitrarily given the code R500.

The following table covers most of the refrigerants in present use.

Refrigerant Designations

Chemical Name	*Chemical Formula*	*Designation*
Carbon/halogen Compounds		
Carbon tetrachloride	CCl_4	R 10
Trichloromonofluoromethane	CCl_3F	R 11
Dichlorodifluoromethane	CCl_2F_2	R 12
Monochlorotrifluoromethane	$CClF_3$	R 13
Carbon tetrafluoride	CF_4	R 14
Dichloromonofluoromethane	$CHCl_2F$	R 21
Monochlorodifluoromethane	$CHClF_2$	R 22
Methylene chloride	CH_2Cl_2	R 30
Methyl chloride	CH_3Cl	R 40
Trichlorotrifluoroethane	CCl_2FCClF_2	R 113
Dichlorotetrafluoroethane	$CClF_2CClF_2$	R 114
Monochlorodifluoromethane and Monochloropentafluoroethane	$CHClF_2/CClF_2—CF_3$ (48.8/51.2% by weight)	R 502
Hydrocarbons		
Methane	CH_4	R 50
Ethane	CH_3CH_3	R 170
Propane	$CH_3CH_2CH_3$	R 290
Butane	$CH_3CH_2CH_2CH_3$	R 600
Isobutane	$CH(CH_3)_3$	R 600a
Ethylene	$CH_2{=}CH_2$	R 1150
Propylene	$CH_3CH{=}CH_2$	R 1270
Inorganic Compounds		
Ammonia	NH_3	R 717
Water	H_2O	R 718
Air	N_2 and O_2	R 729
Carbon dioxide	CO_2	R 744
Sulphur dioxide	SO_2	R 764

Properties of Refrigerants

Reference	Formula	At Atmosph. Pressure		Critical Point		Vapour Pressures in bar (absolute) at Various Temperatures					
		Boils °C	Freezes °C	Temp. °C	Press. –bar	–30°C	–15°C	0°C	30°C	60°C	100°C
R11	$C\,Cl_3\,F$	24	–110	197	43.8	0.07	0.20	0.41	1.24	3.08	
R12	$C\,Cl_2\,F_2$	–31	–158	112	41.3	0.99	1.83	3.09	7.35	15.4	33.1
R13	$C\,Cl\,F_3$	–81	–181	29	38.3	8.72	12.0	20.3	39.0		
R21	$CH\,Cl_2F$	9	–135	172	52	0.17	0.36	0.71	2.15	5.2	
R22	$CH\,Cl\,F_2$	–41	–160	96	48.7	1.54	2.97	5.04	12.1	≃ 24	
R30	$CH_2\,Cl_2$	45	–96	249	46.4	0.03	0.08	0.19	0.69		
R40	$CH_3\,Cl$	–24	–98	143	66	0.76	1.54	2.54	6.53	13.8	
R50	CH_4	–161	–182	–81	46.5						
R113	$CCl_3\,FCCl\,F_2$	87	–35	215	34.1	0.02	0.07	0.15	0.54	1.51	
R114	$CCl\,F\,CClF_2$	3.5	–94	144	32.7	0.21	0.46	0.96			
R170	C_2H_6	–89	–172	32	48.5	11.9	15.9	24.2	45.8		
R290	C_3H_8	–42	–190	94	45.1	1.56	2.88	4.68	10.7	21.3	44
R502	mixture	–46	?	91	41.8		3.52		13.1		
R1150	$CH_2 = CH_2$	–104	–168	9	50	19.6	28.9	40.2			
R717	NH_3	–33	–77	133	113	1.19	2.54	4.27	11.7	26	
R718	H_2O	100	0	374	218	—	—	—	0.04	0.20	1.01
R729	N_2 & O_2	–195	—	–141	37.9						
R744	CO_2	–78	–57	31	73	14.3	22.3	34.9	71		
R764	SO_2	–10	–75	157	78	0.34	0.80	1.5	4.58	10.7	

9. CHIMNEYS

Height: for new stacks use Department of Environment's *Memorandum on Chimney Heights* (Third Edition) 1981, HMSO. It gives simple calculation methods and charts to estimate height based on heat input for boilers or furnaces using clean fuels with less than 0.04% sulphur content and no other noxious chemicals, or for fuels of higher sulphur content based on sulphur dioxide emissions. If adjacent buildings are within 5 times the height first obtained from the charts, or the stack is on, or at the side of, a building further corrections are given in this memorandum if the first height is less than 2.5 times the height of such buildings.

If chimney gases contain noxious fumes or certain chemicals such as chlorine, bromine, etc. in appreciable quantities the "practice notes" issued by the Air Pollution Inspectorate must be consulted.

Gas Plumes: usually rise above top of chimney due to exit velocity and gas temperature if this exceeds ambient. Effect due to velocity usually much smaller than that due to temperature. Prevent unnecessary heat loss from flues and chimney surfaces.

Approx. plume rise due to velocity =

$$\frac{1.456}{1 + \frac{V}{U}} \times \frac{\sqrt{(QU)}}{V} \text{ metres}$$

where V is wind speed m/sec; U is gas exit velocity m/sec; Q is gas volume m^3/sec

Approx. total plume rise due to velocity and temperature =

$$\frac{C\,(M)^{0.25}}{V} \text{ metres}$$

where M is heat content of exit gas mass in MW; C is a constant which actually varies slightly with height of chimney, exit gas velocity and ambient temperature. A typical value is 1070 for exit gas velocities around 15m/s and chimney heights around 30m. Total rise can be corrected for different exit velocities by using first equation to obtain increase or decrease in rise at U values other than 15.

Exit Velocity: try to keep as high as possible to increase plume rise and prevent down-washing of gases which can be obtained in eddies downwind of stack and flow down outside of it if exit velocity too low. Recommended exit velocities (see Chimney Heights memorandum for more details) are:

Up to 2MW heat input – not less than 6m/sec at full load
Above 2MW up to 9MW – not less than 7.5 m/sec at full load
Above 9MW up to 135MW – not less than 15 m/sec at full load

Higher velocities should be used if possible, particularly where plant will operate for most of its life under part-load. Plant relying only on natural draught cannot usually obtain these velocities and values of 3 to 6m/sec are all that can be expected.

At low exit velocities there is a danger of winds causing ambient air penetration some distance down interior of chimney, on upwind side, causing cooling of gases and interior surfaces of chimney. This reduces plume rise and can aggravate condensation problems.

Exit Gases: Disperse as from a point source ($H + P$) above ground where H is chimney height, P is plume rise above chimney top. P reduces as wind speed increases. From the point source the cone of gases develops at approximately 10° angle above and below horizontal above open level ground but other buildings or ground contours can affect this by causing wind eddies. On level ground plume usually touches the ground (point of maximum ground level concentration of gases and small particulate emissions) at: $[4.34 \times (H + P)]^{1.14}$ metres from base of stack – for the most frequent wind speeds in Britain around 15 to 25 times H.

The maximum ground level concentration of SO_2 in parts/100 million is roughly:

$$\frac{104000\,S}{V(H + P)^2}$$

where S is weight per hour in kg of SO_2, passing out of chimney, V is wind speed, m/sec.

This formula shows why plume rise is so important because pollution at ground level depends on the square of $(H + P)$ so even small improvements in plume rise due to increased exit velocity, increased temperature or increased mass flow can decrease glc considerably. Typical design values are for 30pphm SO_2 from a new chimney in an existing industrial area or 50pphm in a clean countryside area.

Natural Draught: few boilers operate today on natural draught only. Forced or induced draught fans, or both, are normal. However, the contribution of a chimney to draught is approx. :

$$3338\,H \left[\frac{1.0}{(T + 273)} - \frac{1.4}{(T_1 + 273)(1 - H/500)}\right] \text{N/m}^2 \text{ (or Pascals)}$$

where H is chimney height in m; T is average gas temperature inside chimney in °C; T_1 is ambient temperature in °C.

Alter the 3338 to 33.38 if answer required in mbar. This formula gives theoretical natural draught as a negative value, i.e. suction or "pull" below atmospheric pressure. In practice losses due to gas friction on flue etc. walls and turbulence at bends, dampers and junctions will reduce the actual draught available at the boiler or furnace.

Position: always try to site a new chimney on the prevailing downwind side of the factory or building.

Number: if at all possible combine gases from several boilers or furnaces into one chimney to give better plume rise. Best results obtained by subdividing chimney into separate flues to maintain exit velocities from working units when others idle. External parts of chimneys should be well insulated to prevent internal condensation.

Corrosion and Acid Dew Points: Any sulphur content in a fuel is converted mainly to SO_2 by combustion, with a small proportion further oxidised to SO_3, dependent on temperature, amount of excess air and certain materials having a catalytic effect. This SO_3 can react with water vapour in the gases and where any internal boiler, flue wall or chimney surfaces are below acid dew point a film of sulphuric acid will condense out and cause corrosion problems with certain materials such as mild steel. Typical acid dew points can range from 120–150°C. SO_2 can give similar condensation problems with sulphurous acid but generally these are far less severe. Any chlorine in the fuel can also cause hydrochloric acid condensation on cooler surfaces. Remedies are (a) operating with higher gas temperatures so that all surfaces are kept above acid dew point, (b) insulating flues and chimneys externally to reduce heat losses and increase internal surface temperatures (c) using certain chemical treatments to neutralise these acids (e.g. injection of a very fine alkaline dust into the flues) (d) using high flue velocities to reduce stagnant gas layers and pockets that might be cooled below the average gas temperatures (e) increasing water temperatures inside boiler water-tubes and heating surfaces to, or above, acid dew point.

Smut Emission: If condensation, particularly acid condensation, occurs in flues and inside chimneys, solid particles, particularly light soot particles, can be trapped on the wet surface and agglomerate in layers until some change in conditions causes them to lift off as flakes, often still containing traces of acid. These can adhere to building roofs, motorcars and clothing and often cause corrosion or staining damage. Measures to eliminate acid condensation will also eliminate smutting. Also examine oil burners to ensure good atomisation, no flame impingement on cool surfaces and complete combustion.

Top of Chimney: as sharp-edged as possible with no overhangs or ornamental parapets that could increase wind eddies or down-draught effects. Existing stacks with low exit velocities can be fitted with cones of fibreglass or corrosion resisting metal to increase velocity but only use a small cone angle to avoid high draught loss. Alternatively stack can be fitted with a full length liner to give increased exit velocity and improved insulation. Never destroy upwards exit velocity by installing chimney top dampers or "chinese hats".

Control of Smoke

The 1956 Clean Air Act is probably best remembered for its very successful attack on smoke from chimneys. It defined "dark smoke" as Shade 2 on the Ringelmann Chart and "black smoke" as Shade 4. Black smoke cannot be emitted for any periods exceeding two minutes in total in any period of thirty minutes and continuous emissions of dark smoke must not exceed four minutes, although there is an allowance for soot blowing. A table has been issued setting out the total permissible duration of dark smoke in any period of eight hours for various combinations of furnaces, with and without soot blowing. As an example, the longest permissible emission of dark smoke, even with four or more furnaces with regular soot blowing is 41 minutes in such an eight hour period.

The 1968 Act closes the loophole of dark smoke emitted other than by a chimney, such as from bonfires but a few exceptions are permissible covered by the Clean Air (Emission of Dark Smoke) (Exemption) Regulations 1969 for such things as the burning of demolition waste, road surfacing material, deceased animal carcasses, contaminated containers, or material burned for fire fighting research at training. It is important to note such exceptions require direct and continuous supervision – such a fire cannot be left to burn itself out.

The Control of Pollution Act 1974 makes it an offence to burn off cable insulation in bonfires. Indeed such cable burning can now only be carried out in premises registered under the provision of the Alkali Act (now Health and Safety at Work Act).

Smoke Control Areas

Local Authorities may designate Smoke Control Areas within which it is an offence to allow any smoke emission from a chimney. For domestic premises this in effect compels the use of gas, electricity, light fuel oils or "smokeless" solid fuels, although certain fireplaces specially designed to burn coal smokelessly are allowable if listed by name in Exempted Fireplace Orders.

These Orders also allow industrial and commercial plant to be operated if equipped with mechanical stokers and if using the correct fuel for which the stoker was designed. The listing of a fireplace or boiler/stoker combination in such Orders is only a permission to install, it is no defence against unauthorised emissions of smoke. Action can still be taken if dark smoke emission occurs.

9. CHIMNEYS

Permissible Emissions of Dark Smoke (Ringelmann Chart Shade 2)

Number of boilers or furnaces to chimney	*Allowable emission in any 8 hour period*	
	Without soot blowing (minutes)	*With soot blowing (minutes)*
1	10	14
2	18	25
3	24	34
4 or more	29	41

Maximum continuous emission of dark smoke – 4 minutes (without blowing)
Maximum emission of black smoke – 2 minutes aggregate in any 30 minutes.

Sampling/Tests of Particulate Emissions

Regulations issued in 1968 enabled a local authority to require an occupier to provide suitable sampling points, access methods and sampling platforms if needed, to his chimney or furnace flues to permit the use of measuring equipment to determine emission levels. One of the methods described in BS 3405 was to be used and the sizes of sampling points and their positions were also to be in accordance with this standard. Six weeks notice had to be given. The Regulation then went on to require the occupier having provided sampling points, on receiving not less than 28 days written notice to make and record such measurements in accordance with the methods described in "Measurement of Solids in Flue Gases" and to report the results within 14 days of the tests to the local authority. While such tests might be required to be repeated the frequency for testing any one chimney should not be less than once in every period of three months unless there were very exceptional circumstances.

At first, under the 1956 Clean Air Act, this regulation applied to all plant using pulverised fuel, or other solid fuel or plants fired with solid matter which burned more than one ton per hour, but they were modified by 1977 regulations when the 1968 Act became law to include smaller plants down to 100 1bs (45.45 kg) or more of solid matter, and liquid or gaseous matter fired plants operating at 1.25 million Btu/hr (1.32 million kJ/hr) or more.

At the same time, occupiers with fairly small plant, using less than one ton per hour of solid matter or less than 28 million Btu/hr (29 million kJ/hr) of liquid or gaseous matter were given the opportunity of serving a "counter-notice" on the local authority, requiring them to carry out the measurements at their expense. Occupiers with larger plant still have to carry out the determinations themselves or engage a third party capable of doing the work with one of the specified types of equipment.

These various Acts and Regulations defined "grit" as being particles exceeding 75 microns in apparent diameter (i.e. not passing through a BS 200 mesh sieve) and "dust" as particles between 75 and 1 micron diameter. Any smaller particulate emission would be classed as "fume". The proportions of grit and dust in the emission are also specified. Grit must not exceed 33% of the emission where furnace maximum continuous output rating does not exceed 16.8 million Btu/hr (17.7 million kJ or 4920 kW) i.e. 16,800 lbs steam/hr (7637 kg/hr) or where furnaces are rated by heat input, 25 million Btu/lb (26.4 million kJ/hr). For plants larger than these figures the permitted grit content must not exceed 20%.

The first point to be made about any measurement associated with smoke or particulate emissions is that there is no point in doing it if, through neglect or other poor adjustment, you know that emissions are bad. Measurements of emissions are only made to verify compliance with regulations.

Your first step should always be to set up the correct fuel/air ratios at the burner and to ensure that the grit arrestor, if you have one, is functioning properly, that the grit hopper has been emptied and, if you have a bag system, that the bags are not damaged.

Grit and Dust Measurement

The methods to be used for measuring grit and dust are set out in British Standard BS3405:1983 *Measurement of Particulate Emissions including Grit and Dust (simplified method)*. They are also described in detail in *Measurement of Solids in Flue Gases* by P.G.W. Hawksley, S. Badzioh and J.H. Blackett (2nd Edition) 1977 published by the Institute of Energy. BS3405 gives details of all calculations and some useful table layouts for presentation of the results at each stage of the test.

BS3405 is stated to be a "simplified method" and indeed there is another method laid down in BS893:1978 Measurement of the concentration of particulate material in ducts carrying gases. Unfortunately this latter is often insisted upon when new plant is installed since it is felt that it gives more accurate results. BS893 defines that many more sampling points should be used across the plane of a flue. This very considerably increases the time taken to carry out sampling and requires the provision of a greater number of access holes.

BS3405 states that measurements of emissions taken in accordance with the laid-down procedure will have an accuracy of + 25% and BS893 states that with the much lengthier procedure accuracy will be + 10%. Bearing in mind these measurements are only to verify that you are within prescribed limits, the difference in degree of accuracy is only important to you if you are close to those limits.

Although a size analysis of the emission is not required under the Acts and measurement Regulations, it is well worth the slight extra costs to carry it out. The cost of preparing the test equipment and using a two man team for the basic emission test is fairly high as several test runs need to be made involving at least one full day site visit. Size analysis then only adds 2% to 3% to the cost.

Collection efficiency is an important consideration. The National Industrial Fuel Efficiency Service (NIFES) has a great deal of experience, through thousands of tests they have carried out, particularly with the BCURA equipment together with a Coulter Counter for particle size analysis.

By far the most commonly used equipment is the cyclone/filter probe developed by the old British Coal Utilisation Research Association (BCURA) and commercially available since 1957. This uses a small cyclone which collects a flue gas sample through a sharp edged inlet nozzle and separates the majority of the grit and dust, which passes into a detachable hopper which can be weighed before and after test. Any very fine dust passes on to a back-up filter system which removes most of this fine material. It is manufactured by Airflow Developments Limited, Lancaster Road, High Wycombe, Bucks HP12 3QP.

When tests are carried out, they must cover "standard" operation of the plant concerned, including any regular peak loads, so that normally several determinations should be carried out and an average hourly emission established. No plant should be said to exceed the permitted emission levels on the results of only a single test run.

Permissible Emission Levels of Grit and Dust

(Readers are referred to Energy Users Research Association Bulletin No.46, of April 1984, which gave a more detailed review of the legislation, methods of measurement etc.)

Two Working Parties were set up by the Department of the Environment. The first examined particulate emissions from boilers and furnaces where the material being heated did not contribute to the emission. From their first report followed the Memorandum *Grit and Dust – the Measurement of emissions/standard levels of emission* – 1967,HMSO and the Clean Air Regulations Nos 161 and 162 of 1971 which enabled local authorities to require sampling points to be provided in chimneys and flues and ask for measurements to be made and results supplied to the authority. The second report has not yet resulted in firm legislation. It recommended maximum emission levels from incinerators, cold blast cupolas and other furnaces where part of the emission was from the material being heated. A draft of a proposed Regulation was circulated in May 1977 which set out the emission figures from the Working Party's Report but this draft has not yet been confirmed. However, anyone considering a new plant of any of these types, or fitting collection equipment to existing plant, should read this report and ensure that his design will meet the emission requirements set out.

The draft Regulation sets lower emission figures for furnaces where the material being heated contributes to the emission than for furnaces in which there is no contribution from the material. Some values for maximum emissions are given in the following two diagrams as a preliminary guide.

Maximum Permissible Emission of Grit and Dust

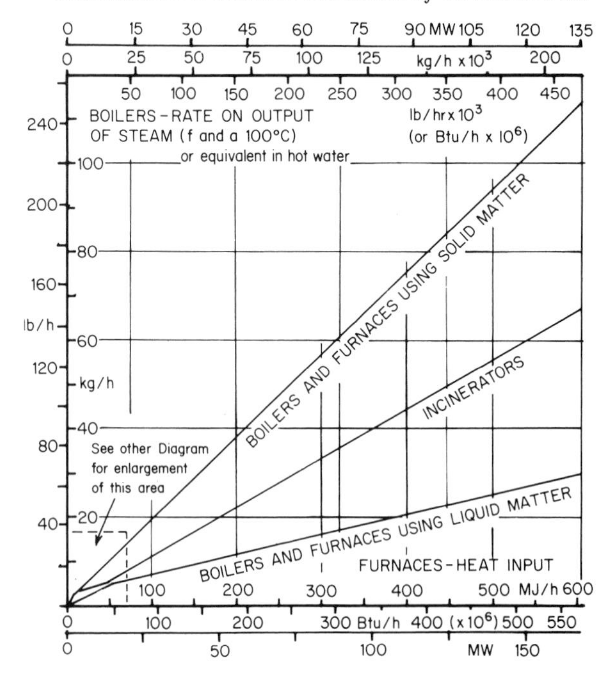

Maximum Permissible Emission of Grit and Dust

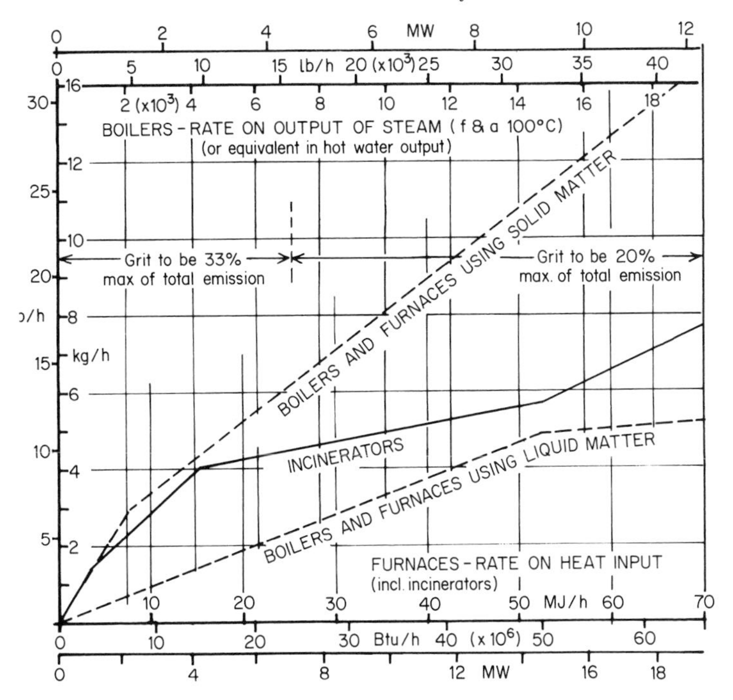

This second diagram is an enlargement of part of the first, showing the rather complicated shape of the three emission lines for small boilers, furnaces and incinerators. The reason is that grit collecting equipment of lower collecting efficiency is permissible for these smaller plants. On both diagrams the "furnace" lines apply only to furnaces where the material being heated does not contribute to the emission. Where the material does contribute, it would be better to take a value from the "incinerator" line by using furnace heat input.

For boilers, the maximum permitted emission is related to the maximum continuous steam or hot water output. For furnaces the emission is related to the heat input. Lower levels are set for liquid (and gaseous) fuel fired plant than for solid fuel. In addition to limits on the total amount of emission the amount of grit (particles not passing through a 200 mesh BS sieve, i.e. above 75 microns diameter) is limited to a minor proportion; 20% of total emission for all medium and large plants, with a relaxation to 33% for smaller plants – as is shown on this second diagram.

These figures of emissions assume that the height of the chimney from which such emissions take place is at least equal to that set by the Memorandum on Chimney Heights (Third Edition) 1981 (HMSO), in order to give a wide dispersion of the emissions.

9. CHIMNEYS

Acts, Regulations, Orders and Other Publications Relating to Air Pollution Through the Combustion of Fuel

1. The Clean Air Act 1956
2. The Clean Air Act 1968
3. The Control of Pollution Act 1974
4. The Health and Safety at Work Act 1974
5. The Clean Air (Measurement of Grit and Dust) Regulations 1968 SI 1968 No.431
6. The Clean Air (Arrestment Plant) (Exemption) Regulations 1969 SI 1969 No.1262
7. The Clean Air (Emission of Dark Smoke) (Exemption) Regulations 1969
8. The Clean Air (Heights of Chimneys) (Prescribed Form) Regulations 1969 SI 1969 No.412
9. The Clean Air (Heights of Chimneys) (Exemption) Regulations 1969 SI No.411
10. The Clean Air (Authorised Fuels) Regulations 1956, 1963, 1965, 1970, (No.2), 1971 (No.3), 1971 (No.4) *et seq.*
11. The Clean Air (Measurement of Grit and Dust) Regulations 1971 SI 1971 No.161
12. The Clean Air (Emission of Grit and Dust) Regulations 1971 SI 1971 No.162
13. Smoke Control Areas (Exempted Fireplaces) Orders 1970 *et seq.*
14. The Dark Smoke (Permitted Periods) Regulations 1958 SI 1958 No.498
15. Draft Regulation – Emission of Grit and Dust from Furnaces 1977 Issued by the Department of the Environment 27th May 1977. Ref.NPCA/366/1. (sets out likely future emission requirements for furnaces, incinerators and driers as recommended by the Second Working Party, using appendices 1–4 of their Report (see item 19). Not on sale but most local authorities have copies which may be inspected.
16. The Fifth report of the Royal Commission on Environmental Pollution. *Air Pollution – An Integrated Approach.*
17. Report of the Working Party on Grit and Dust Emissions 1967.
18. Report of Second Working Party on Grit and Dust Emissions 1974.
19. Grit and Dust – The measurement of emissions from boiler and furnace chimneys Standard levels of emission 1967
20. Chimney Heights – (Third Edition) 1981 – A Memorandum
21. BS2742C *The Ringelmann Chart* 1957
22. BS2742M *The Miniature Smoke Chart* 1960
23. BS3405 1983 *Measurement of Particulate Emissions including Grit and Dust (simplified method)*
24. *Measurement of Solids in Flue Gases* (2nd Edition) 1977 Hawksley et al. Institute of Fuel.

Items 1 to 20 are available from Her Majesty's Stationery Office and Government Bookshops. Items 21 to 23 are available from the British Standards Institution. Item 24 is available from the Institute of Energy, 18 Devonshire Street, London W1.

10. ELECTRICAL

Alternating Currents

Single Phase Power: $W = \frac{V_{ph} \times I \times \text{Cos}\,\phi}{1000}$ kilowatts

Power in balanced three-phase circuit :

$$\frac{\sqrt{3} \times V_L \times I \times \text{Cos}\,\phi}{1000} \text{ kilowatts}$$

$$= \frac{3 \times V_{ph} \times I \times \text{Cos}\,\phi}{1000} \text{kilowatts}$$

Where V_{ph} = Phase voltage (Live to Neutral)
V_L = Line voltage (Live to Live)
I = Current (Line or phase–amps)
$\text{Cos}\,\phi$ = Power factor

Power in an unbalanced 3-phase circuit = $W_1 + W_2 + W_3$

Where W_1, W_2, and W_3 are the wattages in each phase.

Power (kW) may be obtained from a kilowatt-hour meter by timing the rotation of its disc.

$$\text{kW} = \frac{N \times 3600}{K \times T}$$

Where N = No. of revs of disc.
K = Meter constant (rev/kWh)
T = Time in seconds for a given number of revolutions.

The power factor of a single phase circuit can be obtained by connecting a wattmeter and ammeter.

$$PF = \frac{\text{Wattmeter Reading}}{\text{Voltage} \times \text{Current}}$$

The power factor of a balanced three-phase circuit can be obtained by connecting two wattmeters, W_1 and W_2.

The power factor Cos ϕ is obtained from,

$$\tan\,\phi = \frac{3\,(W_1 - W_2)}{(W_1 + W_2)}$$

Where suitable metering exists, an average value of Power Factor over a period can be obtained as follows:-

$$\tan\,\phi = \frac{\text{kilovar-hours}}{\text{kilowatt-hours}}$$

Power Factor Correction

In an electrical AC circuit which contains inductance, such as motors, transformers, discharge lighting, the current drawn from the supply is greater then would be the case with a purely resistive load of the same rating, i.e. electric heating, tungsten lighting.

This is due to the fact that inductive loads require a magnetising current in addition to the actual load current. The magnetising current is 90° out of phase with the load current, and as such, does not impose a kilowatt load on the system. It is known as "wattless" or "idle" current, since it does not contribute to the power used. It does however, result in an increase in the current and consequently the kVA drawn from the supply. Cables, transformers and switchgear have to be rated to accommodate the wattless current, and in addition, most supply authorities impose a cost penalty in respect of excessive kVA demand. It is therefore important to limit the wattless current as much as possible.

The most effective means is by the addition of capacitors into the circuits. The current drawn by a capacitor is wattless (other than small losses) and has the effect of cancelling the magnetising current drawn by inductive loads. Thus although the motor still receives its magnetising current, the current drawn from the supply is minimised, thus reducing the kVA demand, and improving the power factor.

Capacitors may be connected individually at motor terminals or may be installed at motor control boards or main switchboards. In the latter cases, they are generally arranged to be automatically switched in and out in stages as the load increases or decreases.

As an example of the effectiveness of capacitors, consider the case of a 15 kW, 415 Volt, 3-phase induction motor. The full load current of the motor would be 27 amps and the power factor 0.87.

The kVA taken by such a motor would be 19.38 and the current drawn from the system would be 27 amps.

By connecting a suitable capacitor in the motor circuit, the current would be reduced to 24.70 Amps, and the kVA to 17.73.

On the basis of a monthly kVA demand charge of £2.50 this would result in an annual saving of £49.50. The value to which power factor should be improved depends on the supply authority tariff.

Table II gives the current requirements for the British Electricity Boards.

Great care should be taken when installing capacitors on individual motors. If too great correction is fitted, serious damage can be caused due to over-voltages being created at the moment of switching off when the motor may temporarily act as a generator. See BS Code of Practice CP321.102 for detailed recommendations.

Table I shows the amount of capacitance required to improve power factor from a value on the vertical left hand column to a value on the top horizontal line. Read off the figure from this Table, multiply by the actual kW load, and the answer is the kVA of capacitance needed.

Table I. Power Factor Correction

Power factor "as found"	*Size of capacitors in kVAr per kW of load to raise power factor to:*						
	1.0	*0.98*	*0.96*	*0.94*	*0.92*	*0.90*	*0.85*
0.70	1.020	0.811	0.729	0.657	0.591	0.536	0.400
0.72	0.963	0.754	0.672	0.600	0.534	0.479	0.343
0.74	0.909	0.700	0.618	0.546	0.480	0.425	0.289
0.76	0.855	0.652	0.564	0.492	0.426	0.371	0.235
0.78	0.803	0.594	0.512	0.440	0.374	0.319	0.183
0.80	0.750	0.541	0.459	0.387	0.321	0.266	0.130
0.82	0.698	0.489	0.407	0.335	0.269	0.214	0.078
0.84	0.645	0.437	0.355	0.283	0.217	0.162	0.026
0.86	0.593	0.390	0.301	0.230	0.167	0.109	–
0.88	0.538	0.335	0.246	0.175	0.112	0.054	–
0.90	0.484	0.281	0.192	0.121	0.058	–	–
0.92	0.426	0.223	0.134	0.063	–	–	–
0.94	0.363	0.160	0.071	–	–	–	–
0.96	0.292	0.089	–	–	–	–	–

Table II. Current Types of Industrial Tariffs
(These may change in future years)

Area Electricity Board	*Usual Type of Tariff (Industrial)*	*Recommended level of power factor*
London	kW maximum demand	0.85 lag
South Eastern	kW maximum demand	0.85 lag
Southern	kVA maximum demand	0.98 lag
South Western	kW maximum demand with av. power factor clause	0.90 lag
Eastern	kVA maximum demand	0.98 lag
East Midlands	kVA maximum demand	0.98 lag
Midlands	kVA maximum demand	0.98 lag
South Wales	kW maximum demand with av. power factor clause	0.95 lag
Merseyside and N. Wales	kW maximum demand with fixed charge per kvar of maximum demand in excess of 0.4 × kW demand	0.93 lag
Yorkshire	kVA maximum demand	0.98 lag
North Eastern	kW maximum demand with av. power factor adjustment	0.90 lag
North Western	kW maximum demand charge with av. power factor adjustment	0.90 lag
South of Scotland	kW maximum demand with average power factor clause or kVA maximum demand	0.98 lag
North Scotland Hydro	kW maximum demand with av. power factor clause	0.90 lag
Electricity Board N.I.	kVA maximum demand	0.98 lag

Electric Motors

For the vast majority of industrial applications the cage rotor induction motor is used. Of these, the most usual speeds are 3000 rpm (2-pole), 1500 rpm (4-pole), 1000 rpm (6-pole) and 750 rpm (8-pole). The highest efficiencies are obtained at the highest speeds as shown.

Efficiencies of induction motors maximise at full load and fall by about 3 to 5% between full load and half load. Below half load, the efficiency falls rapidly.

The power factor of the cage induction motor also maximises at full load, and falls substantially below 80% full load.

Motor ratings should therefore be chosen so that they operate within 80 to 100% of rating.

Where a motor operates for appreciable periods at less than 50% rating, the use of an electric controller may be advantageous. These have the effect of maintaining a high power factor on low loads, and thus produce a reduction in losses.

The simplest method of starting a cage induction motor is direct-on-the line, in which full line voltage is applied to the motor. This results in a starting current of 5–6 times full load current, and in the smaller installations, this can impose a limit to the rating of the motors which can be started by this means. Alternative methods such as star/delta, wound rotor etc. are available as means of reducing the starting current.

Electronic thyristor controlled starters are now available for motor starting. These limit the starting current to 2–3 times full load current and also eliminate the high starting torque associated with DOL starters which can have an adverse effect on the driven mechanism.

The cage induction motor is essentially a fixed speed machine, the speed being obtained from the expression:

$$N = \frac{f \times 60}{p}$$

Where N = Speed in revs/min.
f = Frequency of the supply (Hertz)
p = Number of pairs of poles

(The actual speed will be slightly less than obtained from the above formula as there is a progressive but small drop in speed as increasing load is applied.)

Variations in speed can be obtained by employing multipole motors which enable various pole arrangements to be selected. This enables the motors to be run at a number of fixed speeds. More recently controllers have been developed which by solid state circuitry allow a standard squirrel-cage motor to be run safely and relaibly at any speed up to about 110% of normal.

Speed variation can also be obtained by varying the rotor resistance. With this type of motor, the cage rotor is replaced by a wound rotor. A variable external resistance is connected to the rotor through slip rings and is used to vary the speed. This type of motor has a low efficiency , and has now largely been replaced by the use of standard cage motors in conjunction with electronic controllers. The controllers produce a variable frequency output which is thus used to vary the motor speed.

Characteristics of Standard Induction Motor

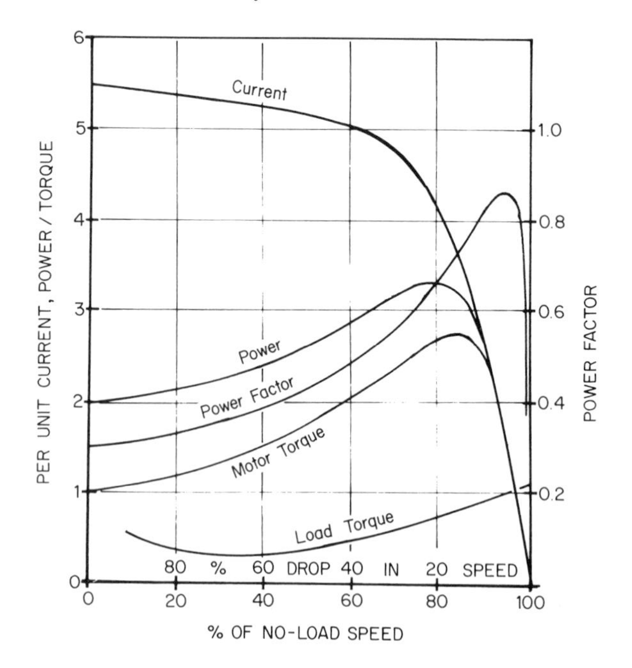

There are two further diagrams on this and the next page. The one on page 92 shows that the efficiency of a motor, particularly in the smaller sizes, improves with speed.

The diagram below shows that the "no-load" current also varies with speed. A 2-pole motor (3000rpm at no-load) typically takes 26% of full load current when running idle, but an 8-pole motor (750rpm at no-load) takes 45%. This difference persists over the whole load range to a diminishing extent – to continue this example, at 40% load the 2-pole motor takes 47%, but the 8-pole takes 58%, of their respective full load currents. Referring back to the diagram on page 92 again the full load currents and power consumptions are lower for the higher speed motors – a 10kW output 2-pole motor would have a full load efficiency of 87%, i.e. power input of 11.5kW, so that its usage at 40% load would be (47% of 11.5) i.e. 5.4kW. The 8-pole 10kW ouput motor's full load efficiency would be 86%, power input 11.6kW and so at 40% load would require (58% of 11.6) = 6.7kW. There is thus a clear saving by using the 2-pole-motor, even if a speed reducing gearbox or belt drive has to be fitted – if this has a 5% power loss, the 5.4kW is only increased to 5.7kW.

These diagrams also demonstrate the need to avoid motors running at low loads all their life. If the driven load requires a motor input of 5kW a 4-pole 7.5kW output motor would take (7.5 × 73%)/86% i.e. 6.4kW, but if an oversized 15kW output motor had been fitted, it would take (15 × 48%)/88% i.e. 8.2kW.

Typical Load/Current Curves for Three-Phase Induction Motors

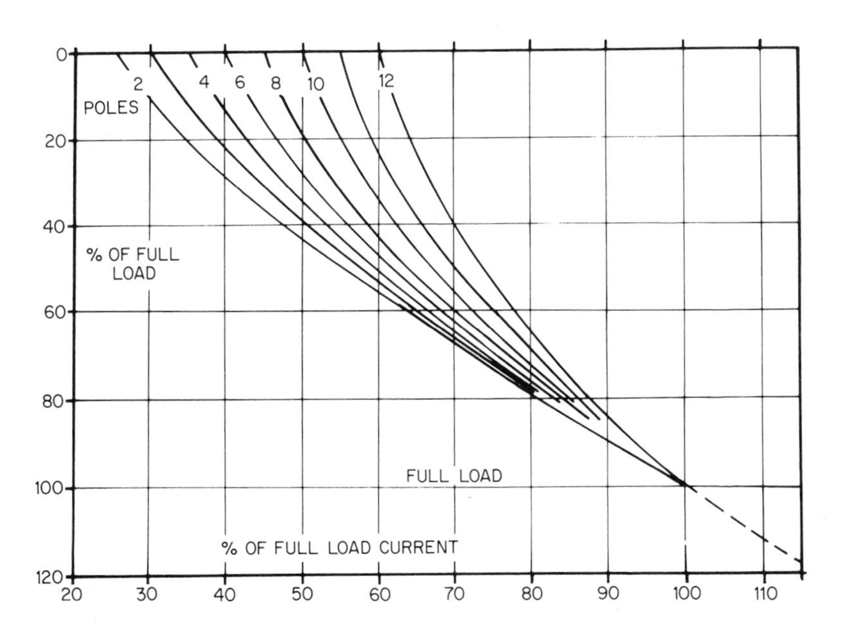

Typical Full Load Efficiencies Three-Phase Induction Motors

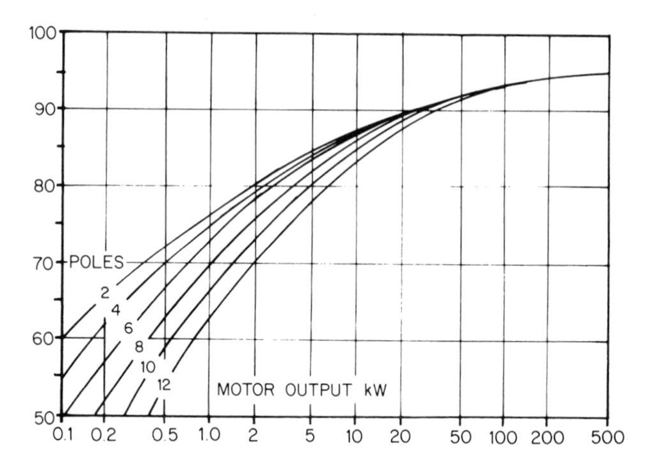

Recommended Maximum Sizes of Capacitors for Fitting on Individual Motors (KVAr)

Motor Output kW	*500rpm (8-pole)*	*1000rpm (6-pole)*	*1500rpm (4-pole)*	*3000rpm (2-pole)*
2	1.6	1.1	1.1	0.6
4	3.8	2.0	1.9	1.3
5	4.0	2.5	2.0	1.5
10	6	4	4	3
20	12	7.5	6	6
30	16	10	10	8
50	24	16.5	14	12
75	32	24	22	18
100	38	28	27.5	23.5
150	54	36	36	32

This Table is compiled from several manufacturer's performance figures for motors. Such sizes of capacitors should improve power factor to over 0.95 for motor loads between 50% and 100% of rating.

Variable Speed Control

Fans, pumps and air compressors are designed to meet a maximum load, which often does not occur, due to excessive margins being allowed for peak loads, or even if it does occur the duration may be small. The normal, lower, flow requirements are usually met by throttling. Fans can have inlet or outlet dampers, pumps a throttle valve, usually on delivery pipe, and compressors may operate on stop/start, or "unloading" by controlling valve gear.

While the squirrel cage motor is usually used as a robust, cheap and reliable power source, efficiency falls at low loads. Also as shown by the diagrams below, the power consumption of a fan or pump does not fall off proportionally if output is reduced by throttling – a pump being particularly bad, requiring almost full load power until throttled below about 40% load.

Modern variable frequency inverter controllers are solid state circuitry to give smooth speed control of an ordinary motor from zero to above normal speed, and can even give a reverse rotation. For motors running continuously at low loads, such controllers can recover their costs in under two years if fitted retrospectively – and can be very attractive if installed with new plant. They will vary speed automatically given a suitable control signal.

Power Requirements of Pumps and Fans at Varying Loads

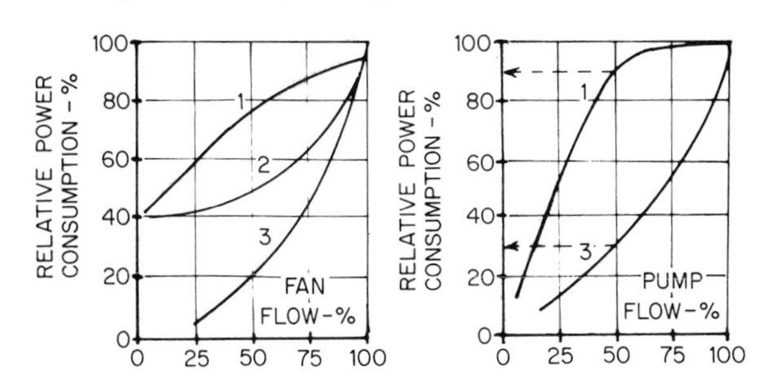

Line 1 on both diagrams shows change in power consumption when flow is throttled by outlet damper or valve.

Line 2 (on fan diagram) shows that inlet vane damper control gives some savings in power consumption compared to outlet damper control.

Line 3 on both diagrams shows the effect of a variable speed controller and the power consumption savings produced compared to lines 1 or 2. For example on a pump running at 50% rating such a controller would save 60% (from 90% to 30%) of power requirement.

Lighting

The provision of adequate lighting in industrial and commercial premises is essential in order to permit the necessary tasks to be carried out in safety and comfort. It is therefore essential that the correct type of lighting be installed in order to give the most satisfactory arrangement compatible with both installation and running costs.

The Table following lists the principal types of luminaires currently available and gives the relevant applications. It also lists the efficiencies of the various lamps in terms of lumen output per watt.

Details of recommended levels of illumination and other information are given in the CIBS Code for Interior Lighting.

The number of luminaires required to provide a given level of illumination over an area can be calculated approximately from the following expression:

$$\text{Lumens} = \frac{\text{Length} \times \text{Breadth} \times \text{Illuminence}}{\text{C of U} \times \text{LLF}}$$

Where C of U = Co-efficient of Utilisation
LLF = Light Loss Factor
The C of U is obtained by first evaluating the Room Index (RI).

$$RI = \frac{\text{Length} \times \text{Breadth}}{\text{Height (Length + Breadth)}}$$

(Height is taken as the height between working plane and the luminaires.)

Having evaluated the Room Index, the C of U is obtained by reference to the manufacturers photometric data sheets.

The light loss factor makes allowance for loss of light output due to deterioration of the lamp, dirt, etc. The value is generally taken as 0.8.

The number of lamps required is obtained by dividing the total Lumens required by the Lumen output of the lamps to be used, i.e.:

$$\text{No. of Lamps} = \frac{\text{Total Lumens}}{\text{Lumens/lamp}}$$

Available Types of Lamp

Lamp Type	*Typical Size (watts)*	*Typical Light Output (lumens /W)*	*Typical Applications*
Tungsten filament bulbs	15 1000	10 20	Domestic, toilets, display lighting
Tungsten Halogen (or halide)	300 2000	19 25	Floodlighting, display, projectors, vehicle headlamps
Mercury blended	100 500	11 28	Becoming obsolescent, being replaced by more efficient types.
Mercury fluorescent	50 2000	31 55	Industrial and road lighting. slight blue colour.
Tubular flourescent – 30/35mm diameter – modern 26mm diameter	4 85 60	15 68 74	Factories, offices shops, schools, hotels. Try to convert to modern 26mm tubes, high efficiency, good colour.
"Bent" fluorescent to fit ordinary lampholder (SL)	18	50	Used instead of low wattage tungsten.
"Bent" fluorescent for small local lighting (PL, 2D, Circolux)	11	63	Life about 5 times longer – same light output for around one-fifth power.
Metal halide	250 2000	70 85	Floodlighting of sports stadium, arenas. High bay industrial lighting.
High Pressure sodium, SON	150 1000	48 110	Medium and high bay industrial, swimming bath, road and area lighting.
Low Pressure sodium	18 180	70 145	Very efficient, road lighting, security and area lighting.
SOX, SOX-E	18 130	70 170	Yellow colour makes it unsuitable for many indoor applications.

(From figures of several lamp manufacturers.)
Light output, lumens/watt, includes current consumption of any control gear.

11. Air, Humidity, Compressed Air and Compressors, Fans

Psychrometry

A psychrometric chart is a graphical representation of the physical properties of humid air; its use can avoid lengthy calculations if only approximate results are required.

The properties normally indicated by the chart are defined:

DB or "dry bulb temperatures"

Actual temperatures of the humid air as registered for example by a thermometer.

DP or "dewpoint temperature"

The temperature at which the air becomes saturated with water vapour, i.e. when actual vapour pressure = saturation vapour pressure.

WB or "wet bulb temperature"

The temperature obtained from a thermometer bulb covered with a moist clean cloth. This may be further sub-divided into "screen wet-bulb" where the thermometer is protected by perforated baffles etc. so that it is standing in still air, and "ventilated wet bulb" where the thermometer is exposed to an appreciable air velocity (sometimes called "sling" wet-bulb).

The reading obtained is lower than DB, and can be used with an appropriate chart or tables to indicate the degree of saturation of the air.

(A "sling" Psychrometer comprises DB and WB thermometers mounted together and whirled to give the air movement.)

Absolute Humidity (Moisture Content)

Weight of water vapour present usually expressed as weight of moisture per unit weight of dry air. Older British charts show grains of water per pound dry air (7000 grains = 1 pound). More modern charts show lb/lb or kg/kg.

Relative Humidity (To Saturation)

At lower temperatures this indicates the percentage of saturation by partial pressure, i.e. the partial pressure of water vapour actually present as a percentage of the partial pressure if the air were saturated with water vapour (at temperatures above water boiling point absolute humidity is often used).

Percentage Humidity

Percentage humidity is sometimes used. This is weight of water vapour actually in air as a percentage of that in saturated air. It is not exactly the same value as relative humidity. Take care not to be confused between the two terms.

Specific Enthalpy (Total Heat Energy)

Heat content consists of latent and sensible components. There is some variation between chart values depending on datum used and on some charts the sensible heat of moisture is not included. This value also is expressed in heat units per unit weight of dry air. On many charts total heat is derived (for simplicity) from lines parallel to the wet bulb lines.

Specific Volume

Usually indicated in terms of unit volume per unit weight of dry air.

$$\text{True specific volume} = \frac{\text{inducted volume}}{1 + \text{weight of moisture present}}$$

The skeleton chart which follows is typical of those available. In the size of this book's pages, only a skeleton can be reproduced legibly. The CIBS publish larger versions and various manufacturers of air conditioning equipment may also include these in their literature. Make sure the datum point is stated.

"Skeleton" Diagram to Illustrate Use of a Psychrometric Chart

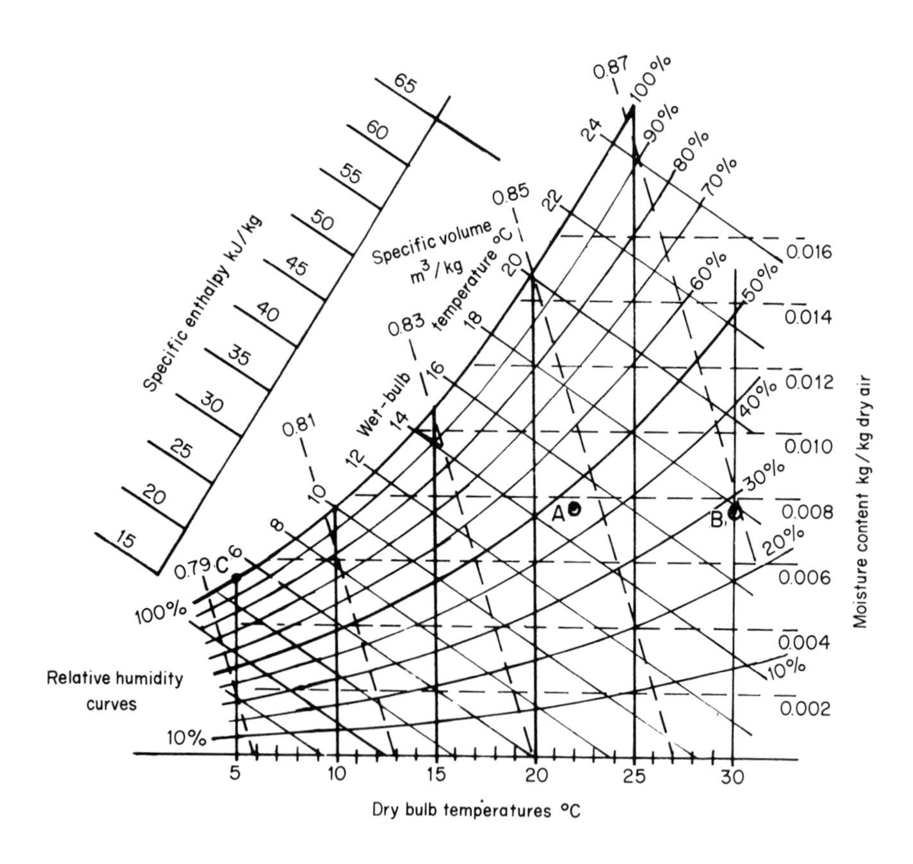

Explanation of Lines on Diagram

Continuous curved lines	–	lines showing percentage saturation or relative humidity of air
Sloping broken lines	–	show specific volume of humid air, as m^3 per kg of dry air, i.e. the volume of 1kg of dry air plus water vapour in the air
Vertical continuous lines	–	show dry-bulb temperatures in °C
Sloping continuous lines	–	show wet-bulb temperatures in °C
Horizontal broken lines	–	show water vapour or moisture content expresssed as kg in 1 kg of dry air.

The scale above and to the left of the diagram shows the specific enthalpy of the humid air, expressed as kJ (of latent and sensible heat of the water vapour associated with 1kg of dry air plus the sensible heat of the dry air) per kg of dry air. On this diagram, the datum point is 0°C, i.e. dry air (zero relative humidity) at 0°C represents zero specific enthalpy. Note that the specific enthalpy lines on the scale are *NOT* parallel to the sloping wet-bulb lines — use a straight-edge from the scale.

Examples of the Diagram

If wet and dry-bulb temperature readings are taken in a current of air and show 22°C DB and 15° WB, what does the diagram tell us?

Find the intersection of the dry bulb and wet bulb lines, marked as point A on the diagram. We can then read off:

Relative Humidity	= 46% (point is between the 40% and 50% curves)
Dew Point	= 10°C (move horizontally from point A to intersect the 100% relative humidity curve and read the dry-bulb temperature vertically below)
Moisture Content	= 0.066 kg/kg dry air (interpolate between the 0.06 and 0.08 horizontal lines)
Specific Enthalpy	= 41.5kJ/kg dry air (using the scale and a straight-edge parallel to the scale lines and passing through A)
Volume	= 0.846m^3/kg dry air (interpolate between the 0.83 and 0.85 sloping lines).

A further use of the diagram is to estimate heat requirements. Suppose it were required to heat this air at point A conditions to 30°C DB? Since moisture/kg dry air remains constant, the air condition will move horizontally to point B, on the 30°C DB line. One can then read:

New Relative Humidity = 28%; New Volume = 0.869 m^3/kg dry air
New Specific Enthalpy = 49.5 kJ/kg dry air, so heat required is 8.0 kJ/kg.

Again if the air at point A conditions were cooled (say by passing over a refrigeration coil) to 5°C what would happen? As the air cooled, move the condition point horizontally to the left until it reaches the 100% RH curve at the dew point of 10°C. On further cooling the air cannot hold all the water vapour as some will condense onto the cold surface, and the condition point will slide down the 100% RH curve until the 5°C DB line is reached at point C. We can then read off:

New Moisture Content = 0.005 kg/kg; New volume = 0.794 m^3/kg
New Specific Enthalpy = 18.5 kJ/kg, so heat removed by cooling = 23.0 kJ/kg.

Diagram to Obtain Relative Humidity of Air from Dry-bulb and Wet-bulb Temperature Readings (wet bulb in airflow)

(This is an enlargement of the first part of the similar diagram 0-140°C.)

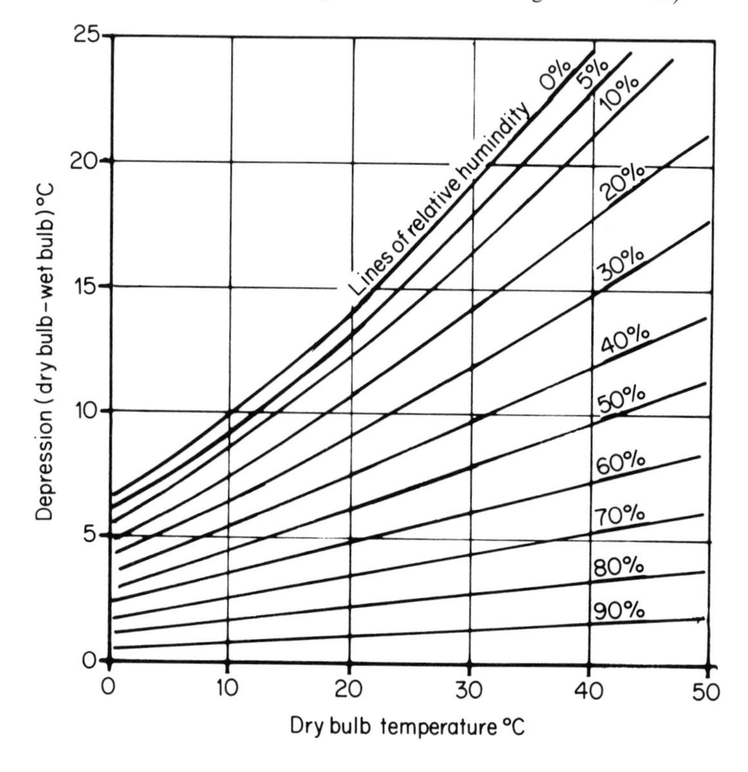

Diagram to Obtain Relative Humidity of Air from Dry-Bulb and Wet-Bulb Temperatures.
Readings (Wet bulb in airflow, not screened)

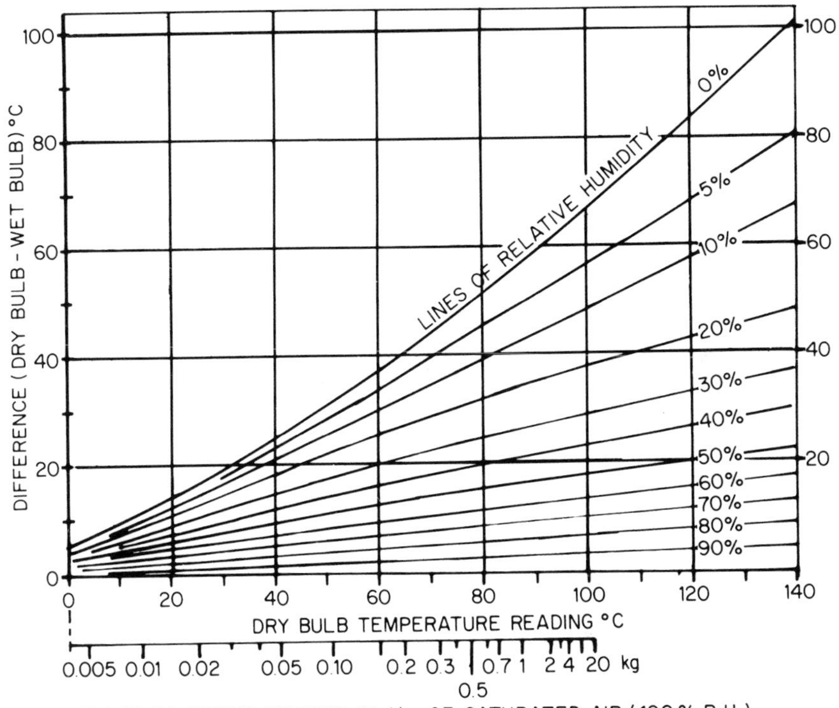

Diagram for Estimation of Heat Gains or Losses due to Air Movement Across a Water Surface at Various Temperatures

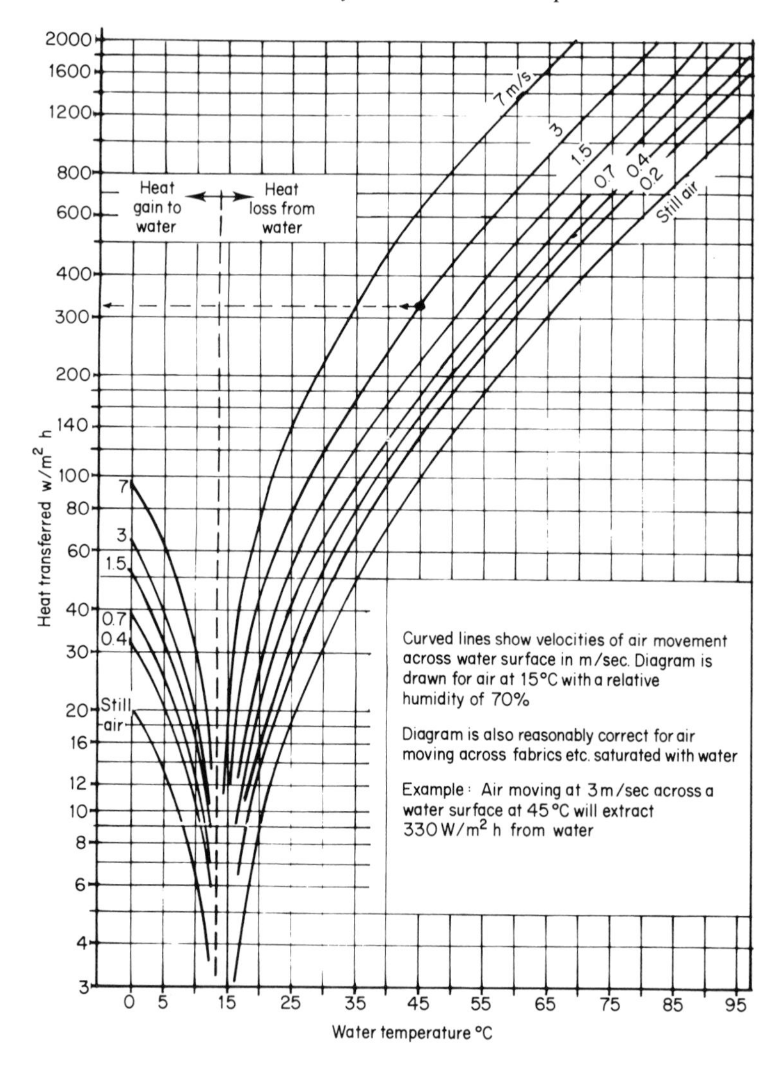

Power Requirements of a Reciprocating Compressor

The power to drive the shaft of such a compressor depends on two factors:

(a) the compression ratio between the absolute pressures, at inlet and at outlet

(b) the "adiabatic expansion exponent" which is the ratio between the specific heat of the gas at constant pressure and the specific heat at constant volume. Values are as follows:

Adiabatic Expansion Exponent for Common Gases (10° – 40°C)

Oxygen, nitrogen, air		Butane	1.10
Hydrogen, Carbon Monoxide	1.40	Ammonia	1.31
Carbon Dioxide	1.28	Steam	1.32
Methane	1.31	Sulphur Dioxide	1.27
Propane	1.15	Chlorine	1.33

Using these values, the diagram enables an estimate to be made of the shaft power to compress any of these gases. An estimate of the power input to a 2 or 3 stage compressor can also be obtained, by treating each stage as a separate machine, with appropriate values of the compression ratio for each. Where the inlet gas temperature differs from 15°C, amend the power obtained from the diagram by the multiplier ($T/288$) where T is actual temperature in K. Remember compression ratios are in absolute pressure terms. Check inlet silencers — excessive baffling may cause considerable pressure drops so that inlet air to compressor may be below atmospheric pressure and increase power requirements.

Power Requirements, Reciprocating Compressors

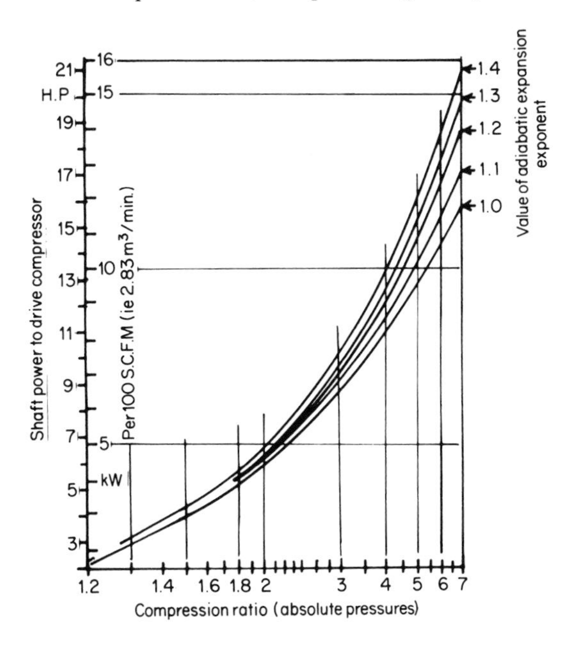

Typical Consumption of Compressed-Air Drills, Tools, etc.

All consumptions in m^3/minute of free air volume at 5.5 bar (approx. 80 psig)

Tool	Size	Consumption
Drills	7mm	0.3 to 0.45
	10mm	0.4 to 0.55
	13mm	0.7 to 0.85
	19mm	1.1 to 1.5
	25mm	1.7 to 2.2
Rotary drives, air per kW output (air motors)	Up to 1kW	1.1 to 1.3
	1 to 4kW	1.1
	Over 4kW	0.9 to 0.95
Torque wrench	7mm	0.3 to 0.4
	13mm	0.7 to 0.95
	25mm	1.1 to 1.5
Spray guns	Small	0.03 to 0.15
	Medium	0.15 to 0.35
	Large	0.35 to 0.7
Blow guns (clearing swarf & dust)		0.15
Sanders		0.3 to 1.2
Screwdrivers		0.2 to 0.7

Assumes tools in good condition with no leaks from joints or couplings. Spray guns usually have reducing valves incorporated.

Approximate efficiencies of reciprocating compressors, for air.

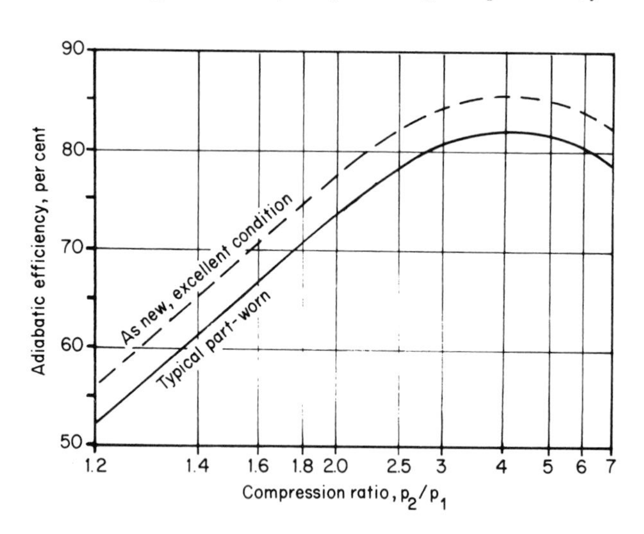

Nomogram for Sizing Delivery Pipes for Compressed Air Systems.

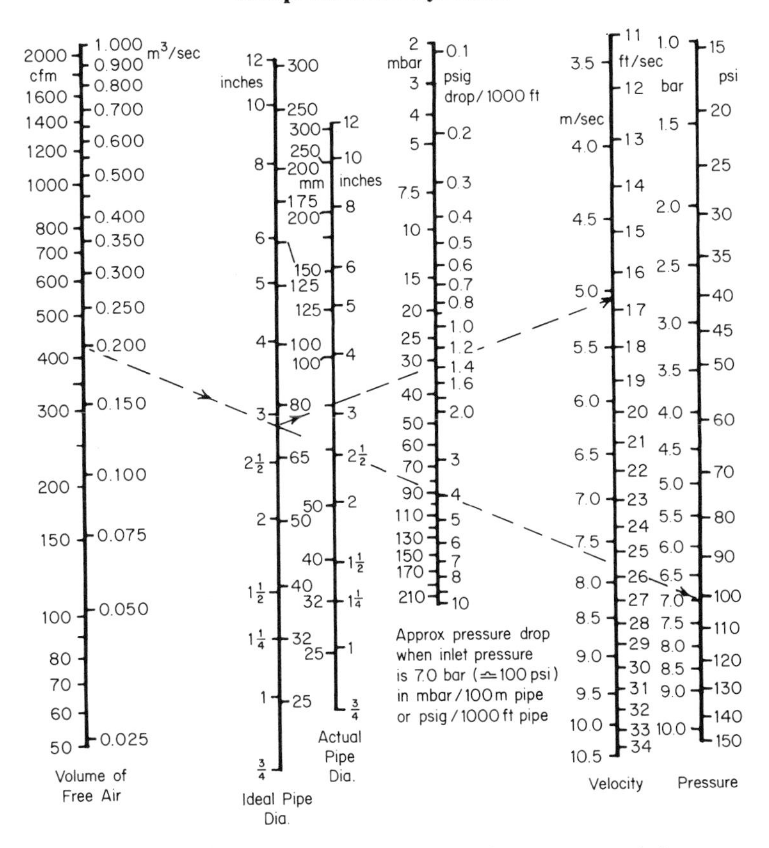

This diagram allows pipe sizes to be chosen to give recommended compressed air velocities of 6m/sec (approx. 20ft/sec). As an example consider an air compressor with an output of 0.2 m³/sec free air (approx 420cfm) at 7 bar (approx 102psi). Place a ruler on 0.2 on the left-hand line and 7 on the right-hand line. This shows as it crosses the second line that the ideal pipe size would be just over 70mm (2.75in) but this is not a standard size. Take the intersection point from this second line to the practical size of pipe actually available, on the third line, say 80mm (or 3″) and project a new line to cross the fourth and fifth lines. This shows an actual pressure drop will be 33 mbar/100m (1.7ps/1000ft) and an actual velocity in such a pipe of 5.2m/sec (18ft/sec). When estimating length of pipe remember to add allowances (as equivalent length of pipe) for bends, tees, valves, etc. Also ensure adequate provision is made for water to be drained off throughout the system. The use of too small a pipe creates excessive pressure losses and so wastes a disproportionate part of the power input to the compressor.

The Fan Laws

For a given system in which the total pressure loss is proportional to the square of the volume flow, the performance of a given fan at any changed speed is obtained by applying the first three rules. The air density is considered unchanged throughout.

Rule 1

The inlet volume varies directly as the fan speed.

Rule 2

The fan total pressure and the fan static pressure vary as the square of the fan speed.

Rule 3

The air power (total or static) and impeller power vary as the cube of the fan speed.

For changes in density:

Rule 4

The fan total pressure, the fan static pressure and the fan power all vary directly as the mass per unit volume of the air which in turn varies directly as the barometric pressure and inversely as the absolute temperature.

For geometrically similar airways and fans operating at constant speed and efficiency the performance shall be obtained by applying the following three rules, and again the air density is considered unchanged throughout.

Rule 5

The inlet volume varies as the cube of the fan size.

Rule 6

The fan total pressure and the fan static pressure vary as the square of the fan size.

Rule 7

The air power (total or static) and impeller power vary as the fifth power of the fan size.

Correction Factors for Fans

Fans are usually rated at sea level and typical ambient air temperatures of say 20°C. For such a fan, delivery will vary with air temperature and pressure (altitude). Delivery varies as barometric pressure in bar; and as (273 + 20) (273 + t) where t is actual temperature °C.

Altitude (Metres)	*Correction*
Sea level	1
200	0.97
400	0.95
600	0.92
800	0.90
1000	0.88
1500	0.83

Temperature (°C)	*Correction*
−10	1.11
0	1.07
10	1.04
20	1
30	0.97
40	0.94
50	0.91
100	0.79

Pressure Loss in Gas and Air Ductwork operated at or near atmospheric pressure

The Monnier formula is:

$$H = \frac{16.62\, d\, Q^2\, L}{D^5}$$

Where H is pressure loss (Pascals); Q is flow rate, m^3/h; L is duct length in m, (with allowances for bends and fittings expressed as equivalent length); D is duct diameter in cm (cm is not a primary SI unit but is used in this formula to avoid an inconvenient value of the numerical constant); d is density of moving gas, relative to air.

If the duct is rectangular, it should be converted to its circular equivalent using the formula:

Equivalent diameter = $2\,WD/(W + D)$ where W and D are the width and depth of the rectangular duct.

This gives the equivalent circular duct in which the gas moves with equal velocity and pressure drop.

Gas (Coal gas or Natural gas) flowing in low pressure mains

Pole's Formula : Quantity $(m^3/h) = 2.254\sqrt{\left(\frac{D^5\, p}{d\, L}\right)}$

Where again D is pipe diameter in cm (chosen to give a convenient numerical constant); p is pressure drop in mbar; d is relative density of gas compared to air; L is length of pipe in m. Allowances must be made for bends and fittings, converting these to equivalent lengths of pipe. Examples being 1.5m for an elbow and 0.6m for a bend.